Reducing Coastal Risk on the East and Gulf Coasts

Committee on U.S. Army Corps of Engineers Water Resources Science, Engineering, and Planning: Coastal Risk Reduction

Water Science and Technology Board

Ocean Studies Board

Division on Earth and Life Studies

NATIONAL RESEARCH COUNCIL
OF THE NATIONAL ACADEMIES

THE NATIONAL ACADEMIES PRESS
Washington, D.C.
www.nap.edu

…DEMIES PRESS 500 Fifth Street, NW Washington, DC 20001

NOTICE: The project that is the subject of this report was approved by the Governing Board of the National Research Council, whose members are drawn from the councils of the National Academy of Sciences, the National Academy of Engineering, and the Institute of Medicine. The members of the committee responsible for the report were chosen for their special competences and with regard for appropriate balance.

Support for this study was provided by U.S. Army Corps of Engineers under contract number W912HQ-09-C-0041. Any opinions, findings, conclusions, or recommendations expressed in this publication are those of the author(s) and do not necessarily reflect the views of the organizations or agencies that provided support for the project.

International Standard Book Number-13: 978-0-309-30586-0
International Standard Book Number-10: 0-309-30586-1
Library of Congress Control Number: 2014948025

Additional copies of this report are available for sale from the National Academies Press, 500 Fifth Street, NW, Keck 360, Washington, DC 20001; (800) 624-6242 or (202) 334-3313; Internet, http://www.nap.edu.

Printed in the United States of America

THE NATIONAL ACADEMIES

Advisers to the Nation on Science, Engineering, and Medicine

The **National Academy of Sciences** is a private, nonprofit, self-perpetuating society of distinguished scholars engaged in scientific and engineering research, dedicated to the furtherance of science and technology and to their use for the general welfare. Upon the authority of the charter granted to it by the Congress in 1863, the Academy has a mandate that requires it to advise the federal government on scientific and technical matters. Dr. Ralph J. Cicerone is president of the National Academy of Sciences.

The **National Academy of Engineering** was established in 1964, under the charter of the National Academy of Sciences, as a parallel organization of outstanding engineers. It is autonomous in its administration and in the selection of its members, sharing with the National Academy of Sciences the responsibility for advising the federal government. The National Academy of Engineering also sponsors engineering programs aimed at meeting national needs, encourages education and research, and recognizes the superior achievements of engineers. Dr. C. D. Mote, Jr., is president of the National Academy of Engineering.

The **Institute of Medicine** was established in 1970 by the National Academy of Sciences to secure the services of eminent members of appropriate professions in the examination of policy matters pertaining to the health of the public. The Institute acts under the responsibility given to the National Academy of Sciences by its congressional charter to be an adviser to the federal government and, upon its own initiative, to identify issues of medical care, research, and education. Dr. Victor J. Dzau is president of the Institute of Medicine.

The **National Research Council** was organized by the National Academy of Sciences in 1916 to associate the broad community of science and technology with the Academy's purposes of furthering knowledge and advising the federal government. Functioning in accordance with general policies determined by the Academy, the Council has become the principal operating agency of both the National Academy of Sciences and the National Academy of Engineering in providing services to the government, the public, and the scientific and engineering communities. The Council is administered jointly by both Academies and the Institute of Medicine. Dr. Ralph J. Cicerone and Dr. C. D. Mote, Jr., are chair and vice chair, respectively, of the National Research Council.

www.national-academies.org

COMMITTEE ON U.S. ARMY CORPS OF ENGINEERS WATER RESOURCES SCIENCE, ENGINEERING, AND PLANNING: COASTAL RISK REDUCTION

RICHARD A. LUETTICH, JR., *Chairman*, University of North Carolina, Chapel Hill
GREGORY B. BAECHER, University of Maryland, College Park
SUSAN B. BELL, University of South Florida, Tampa
PHILLIP R. BERKE, Texas A&M University, College Station
ROSS B. COROTIS, University of Colorado, Boulder
DANIEL T. COX, Oregon State University, Corvallis
ROBERT A. DALRYMPLE, The Johns Hopkins University, Baltimore, Maryland
TONY MACDONALD, Monmouth University, West Long Branch, New Jersey
KARL F. NORDSTROM, Rutgers, New Brunswick, New Jersey
STEPHEN POLASKY, University of Minnesota, St. Paul
SEAN P. POWERS, University of South Alabama, Mobile
DON RESIO, University of North Florida, Jacksonville
AP VAN DONGEREN, Deltares, Rotterdamseweg, The Netherlands

NRC Staff

STEPHANIE E. JOHNSON, Study Director, Water Science and Technology Board
DEBORAH GLICKSON, Senior Program Officer, Ocean Studies Board
ANITA A. HALL, Senior Program Associate, Water Science and Technology Board
SARAH E. BRENNAN, Program Assistant, Water Science and Technology Board

Acknowledgments

Many individuals assisted the committee and the National Research Council staff in their task to create this report. We would like to express our appreciation to the following people who have provided presentations to the committee.

Holly Bamford, National Oceanic and Atmospheric Administration
Robert Banks, U.S. Army Corps of Engineers
Doug Bellomo, Federal Emergency Management Agency
Lynn Bocamazo, U.S. Army Corps of Engineers
Candida Bronson, U.S. Army Corps of Engineers
Bruce Carlson, U.S. Army Corps of Engineers
Craig Colten, Water Institute of the Gulf
Sarah Cooksey, Delaware Coastal Programs
Stephen DeLoach, U.S. Army Corps of Engineers
Gerry Galloway, University of Maryland
John Haines, USGS Coastal and Marine Program Coordinator
Roselle Henn, U.S. Army Corps of Engineers
Carl Hershner, Virginia Institute of Marine Sciences
Keelin Kuipers, National Oceanic and Atmospheric Administration
Michael Lindell, Texas A & M University
David Moser, U.S. Army Corps of Engineers
James Murley, South Florida Regional Planning Council
Michael Park, U.S. Army Corps of Engineers
Lindene Patton, Zurich Insurance Group
Jim Pendergast, U.S. Environmental Protection Agency

Tom Podany, U.S. Army Corps of Engineers
Rhonda Price, Mississippi Department of Marine Resources
Susan Rees, U.S. Army Corps of Engineers
David Rosenblatt, New Jersey Department of Environmental Protection
Tracie Sempier, Mississippi/Alabama Sea Grant
Michael Shelton, Alabama Department of Conservation and Natural Resources
Kate Skaggs, Maryland CoastSmart Communities
Steve Stockton, U.S. Army Corps of Engineers
Robert Young, Western Carolina University
Dan Zarrilli, New York City's Director of Resiliency
Jerome Zeringue, Louisiana Coastal Protection and Restoration Authority

Preface

Coastal regions of the United States are a desirable place to live, work, retire, and recreate. The Atlantic and Gulf of Mexico coasts are home to major population and economic centers, port facilities, and military complexes. Current population growth in southeastern Atlantic and Gulf coastal counties is nearly twice that of the national average. However, these same coasts are subject to impact by some of the most powerful storms on Earth and the destructive potential of these events is increasing due to climate change and relative sea-level rise. High-consequence, low-frequency hazards pose a significant challenge for preemptive decision making because of a lack of personal experience that many have with these events and the probability that an event may not occur during a meaningful time horizon, which may range from a political election cycle to an individual's lifetime. Even though, nationally, we have dealt with significant environmental impact, loss of life, economic devastation, and social disruption from several coastal storms in the past decade, it remains difficult for most coastal residents to fully comprehend the risk of living in these areas. Thus it is challenging for governmental institutions to devote scarce resources to provide protection or forego revenue-generating potential by limiting development in valuable coastal areas to address risk. This behavior is exacerbated when, as a compassionate nation, we rally each time a disaster strikes and provide resources for post-disaster recovery that far exceed those we are willing to provide to manage risk.

The population and economic growth, increase in hazards, unwillingness to proactively manage risk, and pattern of providing substantial post-disaster aid are all contributing to an increase in our risk from coastal natural hazards over time and especially our risk of major impacts due to these events. However, the relatively infrequent nature of coastal natural hazards means that increases in risk today may not manifest themselves in major negative consequences until well into the future. Thus, in many cases we are passing these accumulating disaster-related burdens on to our children and grandchildren.

Given the existing investment, strategic importance, and intrinsic desirability of living in coastal areas, it is unrealistic to believe that we will abandon most of these areas in the foreseeable future. However, living in these areas in a sustainable manner necessitates that we move away from the current disjointed and largely reactive approach to dealing with coastal natural hazards and instead develop a more systematic, proactive approach to managing the risk associated with living in coastal areas.

This study was undertaken as part of a broad 5-year effort to provide advice to the U.S. Army Corps of Engineers on a range of scientific, engineering, and water resources planning issues. Two prior reports issued under this program are *National Water Resources Challenges Facing the U.S. Army Corps of Engineers* (NRC, 2011b) and *Corps of Engineers Water Resources Infrastructure: Deterioration, Investment, or Divestment?* (NRC, 2012a). The current study addresses coastal risk reduction, specifically focusing on reducing flood risks from coastal storm surges along the East and Gulf Coasts. This report and its conclusions are the result of diligent efforts by 13 committee members and 4 National Research Council (NRC) staff representing a diverse range of scientific and engineering expertise. The committee reviewed a large quantity of technical literature; received briefings from multiple federal and state agencies, academic researchers, and members of the private sector (see Acknowledgments); and held lively discussions in meetings that occurred five times over an 8-month period. Three meetings were held in Washington, D.C., one in Mobile, Alabama, and one in Newark, New Jersey. We are particularly indebted to Mr. Bruce Carlson who served as the liaison between the committee and the U.S. Army Corps of Engineers and responded to numerous requests for information and clarification during this study. During the course of the discussions and report preparation, it became clear that assessing, communicating, and managing risk in coastal areas are very challenging concepts even for a committee of experts in coastal science and engineering. I greatly appreciate the time and effort that each committee member invested in trying to understand and synthesize this complex issue and the collegiality, patience, and good humor that members exhibited throughout.

The committee and, particularly, the committee chair are extremely grateful to the NRC staff who supported this study: Stephanie Johnson, Study Director; Deborah Glickson, Senior Program Officer; Anita Hall, Senior Program Associate; and Sarah Brennan, Program Assistant. Stephanie orchestrated the study for the NRC, which was especially challenging given its rapid time line and the diverse set of issues that were involved. Her tenacity, deftness finding information and references, skill separating "wheat from chaff," ability to synthesize complex subjects, and management style were outstanding. Deborah provided a very helpful complement to Stephanie in terms of her perspective on the issues and help shouldering the load associated with this accelerated study. Anita provided excellent administrative and logistical support for the meetings and production of the final report, with assistance near the end of the study from Sarah. This report would not have been possible without their collective skills and extensive efforts; I know the entire committee joins me in expressing our profound appreciation for their contributions.

This report was reviewed in draft form by individuals chosen for their breadth of perspectives and technical expertise in accordance with the procedures approved by the National Academies' Report Review Committee. The purpose of this independent review was to provide candid and critical comments to assist the institution in ensuring that its published report is scientifically credible and that it meets institutional standards for objectivity, evidence, and responsiveness to the study charge. The reviewer comments and draft manuscript remain confidential to protect the deliberative process. We thank the following reviewers for their helpful suggestions, all of which were considered and many of which were wholly or partly incorporated in the final report: Brian Atwater, University of Washington; Michael Beck, The Nature Conservancy; Rudolph Bonaparte (NAE), Geosyntec Consultants, Inc.; Robin Dillon-Merrill, Georgetown University; Jenifer Dugan, University of California; Billy Edge, North Carolina State University; Gerald Galloway, University of Maryland, College Park; Charles Groat, The Water Institute of the Gulf; Jennifer Irish, Virginia Tech; Jim Johnson, Independent Consultant, Columbia, MD; Sandra Knight, University of Maryland, College Park; Mark Mauriello, Edgewood Properties; and Adam Rose, University of Southern California.

Although these reviewers provided many constructive comments and suggestions, they were not asked to endorse the conclusions and recommendations nor did they see the final draft of the report before its release. The review of this report was overseen by John Boland, Johns Hopkins University and Michael Goodchild, University of California, Santa Barbara. Appointed by the NRC, they were responsible for making certain that an independent examination of this report was carried out in accordance with institutional procedures and that all review comments received full consideration. Responsibility for the final content of this report rests entirely with the authoring committee and the NRC.

Richard A. Luettich, Jr., *Chair*
Committee on U.S. Army Corps of Engineers Water Resources Science, Engineering, and Planning: Coastal Risk Reduction

Contents

Summary

Hurricane- and coastal-storm-related economic losses have increased substantially over the past century, largely due to expanding population and development in the most susceptible coastal areas. Eight U.S. cities (Miami, the New York-Newark region, New Orleans, Tampa-St. Petersburg, Boston, Philadelphia, Virginia Beach, and Baltimore) rank among the world's top 20 in terms of estimated potential average annual losses from coastal flooding. Hurricanes Sandy (2012) and Katrina (2005) recently raised awareness of this vulnerability. Climate change poses additional threats to coastal communities. Climate projections suggest possible increases in the strength and frequency of the most intense hurricanes, and sea-level rise will increase the likelihood of major flood events.

Concurrent with the growth in economic losses from natural hazard, there has also been a substantial shift in the source of funds used to cover these losses in the United States. Over the past 60 years, the federal government has assumed an increasing proportion of the financial responsibility associated with coastal storms. This trend highlights the challenges ahead, particularly if federal post-disaster relief discourages state and local governments from taking appropriate actions to reduce risk and enhance resilience.

A wide array of strategies exists for managing coastal storm risks. One set of strategies aims to reduce the probability of flooding or wave impact. These include hard structures, such as seawalls, levees, flood walls, and storm surge barriers, and nature-based risk reduction strategies, such as beach nourishment, dune building, and restoration or expansion of natural areas, such as oyster reefs, salt marshes, and mangroves.

Coastal risk is defined in this report as the potential for coastal storm hazards, such as storm surge–induced flooding and wave attack, to cause adverse effects on human health and well-being; economic conditions; social, environmental, and cultural resources; infrastructure; and the services provided within a community.

Another set of strategies aims to reduce the number of people or structures in areas at risk or to make them less vulnerable to coastal storms. These include design strategies, such as elevating or floodproofing buildings, and "nonstructural strategies," such as relocation and land-use planning to steer future development or redevelopment away from high hazard areas. Over the past century, most coastal risk management programs have emphasized coastal armoring, while doing little to decrease development in harm's way.

This study was undertaken as part of a broad five-year effort to provide advice to the U.S. Army Corps of Engineers (USACE) on a range of scientific, engineering, and water resources planning issues. It examines risk reduction strategies to address coastal storms (hurricanes, tropical storms, and extratropical storms) and associated storm surges, focusing on the East and Gulf Coasts where large coastal storms predominantly occur, and the report outlines principles to guide future U.S. investments in such strategies (see Box S-1 for the statement of task). Other coastal hazards, such as erosion from mild or moderate storms, wind damage, or tsunami-induced flooding, are not considered in depth.

This report calls for the development of a national vision for managing risks from coastal storms (hereafter, termed "coastal risk") that includes a long-term view, regional solutions, and recognition of the full array of economic, social, environmental, and life-safety benefits that come from risk reduction efforts. To support this vision, a national coastal risk assessment is needed to identify those areas with the greatest risks that are high priorities for risk reduction efforts. Benefit-cost analysis, constrained by other important environmental, social, and life-safety factors, provides a reasonable framework for evaluating national investments in coastal risk reduction. However, extensive collaboration and additional policy changes will be necessary to fully embrace this vision and move from a nation that is primarily reactive to coastal disasters to one that invests wisely in coastal risk reduction and builds resilience among coastal communities.

BOX S-1
Statement of Task

The National Research Council's Committee on U.S. Army Corps of Engineers (USACE) Water Resources Science, Engineering, and Planning: Coastal Risk Reduction was assembled to provide advice on reducing flood risks from coastal storm surges along the East and Gulf Coasts. The committee was tasked to address the following questions:

1. What coastal risk-reduction strategies have been used along the U.S. East and Gulf Coasts to reduce impacts of coastal flooding associated with storm surges, and what design standards or levels of protection have been used? To what extent have these many strategies and levels of protection proven effective in terms of economic return, protection of life safety, and minimizing environmental effects?
2. What are the regional and national implications of expanding the extent and levels of coastal storm surge protection? Examples might include operations and maintenance costs, sediment availability, and regional-scale sediment dynamics.
3. How might risk-related principles contribute to the development of design standards for coastal risk-reduction projects? How might risk-related principles increase the ability of coastal regions and communities to prepare for coastal storms and surge, and adjust to changing coastal dynamics, such as prospects of sea level rise?
4. What general principles might be used to guide future investments in U.S. coastal risk reduction?

LANDSCAPE FOR COASTAL RISK MANAGEMENT

The committee's review of the institutional landscape as it relates to federal, state, and local coastal risk reduction efforts (Chapter 2) resulted in the following conclusions.

Responsibilities for coastal risk reduction are spread over a number of federal, state, and local agencies, with no central leadership or unified vision. Multiple federal agencies play some role in coastal risk management, and each agency is driven by different objectives and authorities. No federal coordinating body exists with the singular focus of mitigating coastal risk, although several efforts are under way to increase coordination.

The lack of alignment of risk, reward, resources, and responsibility as it relates to coastal risk management leads to inefficiencies and inappropriate incentives that serve to increase the nation's exposure to risk. Developers, builders, and state and local governments reap the rewards of coastal development but do not bear equivalent risk, because the federal government has borne an increasing share of the costs of coastal disasters. The resulting moral hazard leads to continued development and redevelopment in high-hazard areas.

The vast majority of the funding for coastal risk-related issues is provided only after a disaster occurs, through emergency supplemental appropriations. Pre-disaster funding for mitigation, preparedness, and planning is limited, and virtually no attention has been given to prioritization of coastal risk reduction expenditures at a regional or national scale to better prepare for future disasters. Thus, efforts to date have been largely reactive and mostly focused on local risks, rather than proactive with a regional or national perspective. Also, although the federal government encourages improved community resilience, only a small fraction of post-disaster funds are specifically targeted toward mitigation efforts.

Few comprehensive regional evaluations of coastal risk have been performed, and the USACE has no existing institutional authority to address coastal risk at a regional or national scale. Given the enormous and rising cost of coastal disasters within the United States, improved systemwide coastal risk management is a critical need within the federal government. Under the current planning framework, the USACE responds to requests at a local level on a project-by-project basis, and several major urban areas remain at significant risk. Congressional authorization and funding would be needed for the USACE to undertake a comprehensive national analysis of coastal risks.

PERFORMANCE OF RISK REDUCTION STRATEGIES

Chapter 3 reviews what is known regarding the proven performance, costs, and benefits of hard structures and nature-based strategies to reduce the hazards (e.g., flooding, wave attack) associated with coastal storms and nonstructural and building design measures to reduce the consequences of coastal hazards. Determination of the optimal coastal risk reduction will be site-specific and dependent on an analysis of long-term costs, benefits, and environmental impacts and may involve multiple approaches implemented together.

Beach nourishment and dune-building projects for coastal risk reduction can be designed to provide increased ecological value. Beachfill projects provide some level of risk reduction for coastal infrastructure from erosion, flooding, and wave attack and may reduce the likelihood of forming new inlets. Beach nourishment and dune building do not, however, address back-bay flooding, which may be better addressed by structural measures on the bay side. The short-term environmental impacts of nourishment projects on biological communities is significant, and long-term cumulative ecological implications remain unknown because of the difficulty and cost of mounting large scale monitoring projects and the limited time frame of existing studies. Coastal systems can be managed for multiple uses and benefits, although some compromises may be necessary to optimize benefits across a range of objectives. Improvements for ecological benefits of beach nourishment and dune construction would involve different design specifications that are unlikely to greatly increase construction costs, although they may require alternative approaches to post-construction beach and dune management.

Sediment management should be viewed on a regional basis, rather than on a project-by-project basis. Federal and state agencies have documented plentiful offshore sand deposits for beach nourishment, but not all are of optimal quality or conveniently located to project needs, which could increase costs. Coastal projects can minimize sediment losses by retaining dredge material or emphasizing reuse, as in sand backpassing or bypassing operations. Use of a sediment source that is compatible with a beachfill project site also decreases ecosystem recovery time and enhances habitat value in the nourished area.

Conservation or restoration of ecosystem features such as salt marshes, mangroves, coral reefs, and oyster reefs provides substantial ecological benefits and some level of risk reduction against coastal storms, but the risk reduction benefits remain poorly quantified. Coastal habitats provide numerous ecosystem services, including carbon sequestration, improved water quality, and essential habitat for fish species targeted by commercial and recreational fisheries. Much is known about the capacity of nature-based features to reduce coastal erosion from smaller storms, but additional research is needed to better understand and quantify the effects of natural features (other than beaches and dunes) on storm surge, wave energy, and floodwater inundation. In general, the level of risk reduction provided by oyster reefs and seagrasses appears much lower than that provided by constructed dunes and hard structures, and most of the benefits are associated with reductions in wave energy during low- to moderate-energy events. Research has documented reductions in peak water levels from salt marshes and mangroves, but certain storm conditions and large expanses of habitat are needed for these to be most effective. Thus, many of these nature-based alternatives can only be used for coastal risk reduction at locations that have sufficient space between developed areas and the coastline. Additional quantitative modeling and field observation are needed to better understand and quantify the efficacy of nature-based approaches for coastal risk reduction.

Hard structures are likely to become increasingly important to reduce coastal risk in densely populated urban areas. Many large coastal cities lack the space necessary to take advantage of nature-based risk reduction approaches alone and will instead need additional hard structures to substantially reduce coastal hazards. Adverse environmental impacts commonly accompany the construction of hard structures, although modified designs are possible to reduce these effects. Coupling nature-based approaches with hard structures to buffer the structures against wave attack provides an effective coastal risk reduction strategy if space allows.

Strategies that reduce the consequences of coastal storms, such as hazard zoning, building elevation, land purchase, and setbacks, have high documented benefit-cost ratios, but they are given less attention by the federal government and are viewed as difficult to implement by states. Studies have reported benefit-cost ratios between 5:1 and 8:1 for nonstructural and design strategies that reduce the consequences of flooding, but between 2004 and 2012, federal funds for such strategies were only about 5 percent of disaster relief funds. Those nonstructural and design strategies that are commonly implemented, such as public information campaigns and elevation of in situ development, tend to avoid property rights issues, do not threaten economic interests, and do not generate political opposition.

PRINCIPLES FOR GUIDING INVESTMENTS IN COASTAL RISK REDUCTION

Investments in coastal risk reduction generate significant benefits to society by reducing risk to people and property, but they also involve significant costs. Chapter 4 reviews two approaches for determining what investments in coastal risk reduction are justified: (1) a risk-standard approach and (2) a benefit-cost approach. The *risk-standard approach* recommends investments in coastal risk reduction measures to achieve an acceptable level of risk reduction, and develops cost-effective strategies to meet this level. The *benefit-cost approach* recommends investment in coastal risk reduction when the benefits of the investment exceed the costs. Thus, the level of risk reduction provided by projects under a benefit-cost approach could vary widely based on the costs and benefits provided. While each approach has considerable appeal, each also has at least one significant weakness. For the risk-standard approach it is difficult to factor in non-risk-related benefits or costs, such as environmental benefits. In the case of the benefit-cost approach, it is difficult to evaluate all environmental and social impacts in monetary terms. Given the limitations with each approach, there are advantages of not rigidly adhering to either approach in its purest form but instead incorporating some elements from each.

Benefit-cost analysis constrained by acceptable risk and social and environmental dimensions provides a reasonable framework for evaluating coastal risk management investments. Investments in coastal risk reduction should be informed by net benefits, which include traditional risk reduction benefits (e.g., reduced structural damages, reduced economic disruption) and other benefits (e.g., life-safety, social, and environmental benefits), minus the costs of investment in risk reduction and environmental costs. However, because it is difficult to quantify and monetize some benefits and costs, it is important to expand the analysis to include considerations of difficult-to-measure benefits or costs through constraints on what is considered acceptable in social, environmental, and risk reduction dimensions. Such unacceptable levels of risk may include a level of individual risk of fatality, the risk of a large number of deaths from a single event, or adverse impacts on social and environmental conditions that may be difficult to quantify in monetary terms. It is difficult, however, to establish societally acceptable risk standards and requires extensive stakeholder engagement. Setting such a standard requires value judgments, on which not all individuals or groups will necessarily agree.

The recently updated federal guidance for water resources planning—the 2013 *Principles and Requirements for Federal Investments in Water Resources*[1]—provide an effective framework to account for life safety, social impacts, and environmental costs and benefits in coastal risk reduction decisions. The *Principles and Requirements*, developed by the White House Council on Environmental Quality in response to a 2007 congressional mandate, represent the first steps toward federal water resources policy reform. The document, which applies to water resources investment decision making across the federal government—not just within the USACE—recognizes that water resources investment decisions should also consider social and environmental impacts and not give primacy to benefits or costs that are easily measurable in monetary terms. This represents a significant improvement upon current USACE planning, which uses separate accounts for social and environmental impacts, with largely qualitative measures, effectively relegating such considerations to second-class status behind net economic benefits. Progress has been made on measuring improvements in economic terms and on measuring the value of some ecosystem services and social benefits. For other environmental and social factors that are not easily measured in dollar terms, the *Principles and Requirements* recognize that these costs and benefits should also be given adequate weight in decision-making. **The Council on Environmental Quality should expedite efforts to complete the detailed accompanying guidelines for implementing the 2013 *Principles and Requirements*, which are required before this framework can to be put into action to improve water resources planning and coastal risk management decision making at the federal agency level.**

Until the updated guidelines to the *Principles and Requirements* are finalized, there are steps the USACE could take to improve consideration of multiple benefits and costs in the current decision process. Specifically, further attempts in the USACE planning process could be made to more

---

[1]See CEQ (2013).

quantitatively consider information about social and environmental effects. For example, work that has been done on how to value ecosystem services could be used to value some environmental quality benefits. Once quantified, these costs and benefits should be rigorously considered and clearly communicated to stakeholders. Such an approach could result in different decision outcomes if the additional social and environmental benefits make certain strategies more acceptable to local sponsors and stakeholders than others. However, trying to quantify or monetize social effects and some environmental effects remains challenging. When only some benefits or costs are monetized, there is a tendency to overlook or downplay nonmonetized benefits or costs, and additional attention and/or institutional mechanisms are needed to ensure that these benefits are given adequate weight.

There is no solid basis of evidence to justify a default 1 percent annual chance (100-year) design level of coastal risk reduction. The 100-year flood criterion used in the National Flood Insurance Program was established for management purposes, not to achieve an optimal balance between risk and benefits. There is also no evidence that reducing risk to a 1 percent annual-chance event is in the best interests of society or that this level is necessarily acceptable to the general public. This level of risk reduction may be appropriate in some settings, unwarranted or excessive in others, and inadequate in highly developed urban areas. Such decisions should, instead, be informed by risk-constrained benefit-cost analyses reflecting site-specific conditions.

VISION TOWARD COASTAL RISK REDUCTION

To address the rising costs of coastal disasters, increasing coastal risks in the context of climate change, and the fragmented risk management framework, Chapter 5 presents the committee's recommendations for reducing the nation's coastal risks.

A national vision for coastal risk management is needed in order to achieve comprehensive coastal risk reduction. Effective coastal risk management for the United States requires a national perspective to achieve the most benefits from federal investments and regional solutions, rather than piecemeal, project-by-project approaches. Coastal risk management requires a long-term vision, recognition of the wide array of potential benefits, and coordination of efforts that are currently spread across many agencies that sometimes operate under conflicting mandates. Developing and implementing a national vision for coastal risk management is not the responsibility of any single agency alone, but will require federal leadership and extensive collaboration among federal, state, and local agencies.

The federal government, working closely with states, should establish national objectives and metrics of coastal risk reduction. Specific metrics for coastal risk management could be used by state and local governments to identify necessary actions and assess progress.

The federal government should work with states to develop a national coastal risk assessment. The geographic patterns of disaster risk represented by human fatalities, economic losses, and social impacts can illustrate where the risks are greatest and in need of targeted risk reduction interventions. This analysis should not be based merely on the recent history of hazards but on a comprehensive assessment of risk, including multiple types of hazards under current and anticipated future conditions. The results of the risk assessment would serve as a powerful communication tool for the public and local and national decision makers. The national interest in coastal risk reduction may vary from one community to another, but this would not preclude a community from investing in risk reduction efforts. The risk assessment would serve as a basis to assess the economic, life-safety, social, and environmental costs and benefits under various risk management scenarios, although additional model development is needed to fully support such an effort.

Stronger incentives are needed to improve pre-disaster risk management planning and mitigation efforts at the local level. Hazard mitigation and adaptation planning has significant potential to reduce coastal risk, but most state and local mitigation plans are currently poor and give limited attention to land-use strategies. In light of behavioral and cognitive factors associated with low-probability, high-consequence events, additional focused efforts and stronger incentives (or disincentives for inaction) are necessary to improve the quality of these plans and the breadth of nonstructural mitigation strategies considered. For example, the federal government could adjust USACE cost sharing for coastal risk reduction projects according to the extent and quality of local hazard mitigation planning and the degree to which mitigation is incorporated into other local planning efforts (e.g., land use, transportation). The potential for strategic incentives to improve development decisions or facilitate retreat should be carefully examined in the context of long-term cost savings. Federal and state governments should also work to build commitment to coastal risk reduction among stakeholders and local officials.

The USACE should seize opportunities within its existing authorities to strengthen coastal risk reduction. Although the USACE is limited in its capacity to independently initiate national coastal risk reduction strategies under its current authorities, it can use its existing planning framework to rigorously account for social and environmental costs and benefits, thereby supporting a more holistic view of coastal risk management. Additionally, the USACE should increase incentives for sound coastal planning and continue to develop and improve modeling tools to support state and local planning efforts. The USACE should also look for opportunities to apply adaptive management to enhance learning and improve coastal risk reduction strategies. The USACE should reevaluate its typical 50-year planning horizon and consider longer-term planning in the context of projected increases in sea level to assess the adaptability and long-term costs and benefits (including social and environmental effects) associated with risk reduction alternatives.

1

Introduction

Hurricane Sandy heightened the nation's awareness of the vulnerability of coastal areas to hurricane damage. Eight U.S. cities (Miami, the New York-Newark region, New Orleans, Tampa-St. Petersburg, Boston, Philadelphia, Virginia Beach-Norfolk, and Baltimore) are among the top 20 cities in the world at risk from coastal storms, based on an estimate of potential average annual flood loss of valuable assets (e.g., buildings, transportation, utilities, personal property) (Hanson et al., 2011; Hallegatte et al., 2013). Other large cities along the East and Gulf Coasts, such as Houston, Texas, and countless smaller cities and developed areas, are also vulnerable to coastal storms. New York, New Orleans, and Miami were poorly prepared for a major storm as shown by Hurricanes Sandy (2012), Katrina (2005), and Andrew (1992). If not adequately prepared, coastal cities and developed areas are extremely vulnerable to hurricanes, which can leave many thousands of people homeless, cause extensive property damage, and result in short- and long-term economic disruptions. This chapter provides an introduction to the coastal storm-related risks (hereafter, termed "coastal risks") faced along the U.S. East and Gulf Coasts, and discusses how those risks have changed and are continuing to change. General strategies that can reduce risk and help make communities more resilient to coastal storms are also discussed.

NATIONAL DISASTERS AND COASTAL RISK

The United States has experienced extensive and growing loss from natural disasters. Dollar losses due to tropical storms and floods have tripled over the past 50 years (accounting for inflation; Gall et al., 2011[1]) and currently comprise approximately half of all natural disaster losses (Table 1-1, Figure 1-1). In addition to growth in absolute dollars, per capita natural disaster losses have also grown as have losses normalized by income, highlighting the growth in their relative economic impact (Gall et al., 2011). Appendix A provides a table of major coastal storms that have struck the United States since 1900—most of which made landfall on either the East or Gulf Coasts.

From 1980 to 2013, there were 151 weather- or climate-related natural disasters that caused a direct economic impact on the United States of greater than 1 billion dollars (in 2013 dollars).[2] Tropical cyclones (including tropical storms and hurricanes) compose the single largest category, accounting for 33 of the events (or 22 percent) and 49 percent of the total damage (Table 1-1). During this period, when averaging over 5-year periods, tropical cyclone events causing billion-dollar losses increased from approximately 0.4 per year to over 1 per year, and the losses increased from approximately $1.75 billion per year to as high as $45 billion per year in the 5-year span that includes Hurricane Katrina (Figure 1-1). This increase follows a much more gradual upward trend in tropical cyclone–related economic losses extending back to at least 1900 (Pielke et al., 2008).

Causes of Increasing Disaster Losses

There are two primary reasons for the dramatic increase in natural disaster–related losses: an increase in the people and property in harm's way and an increase in the frequency or severity of the hazard events. Pielke et al. (2008) concluded that growth in tropical cyclone–related economic losses in the United States since 1900 has been minimally influenced by changes in storm climatology and rather is primarily explained by the movement of people and accompanying wealth to areas that are at higher risk. From the east coast of Florida through the Gulf Coast, population

[1]Gall et al. (2011) data included direct loss estimates from the Special Hazard Events and Losses Database for the United States (SHELDUS), federal individual and public assistance and some Hazard Mitigation Grant Program spending associated with presidential disaster declarations, National Flood Insurance Program claims, and privately insured hazard claims.

[2]See http://www.ncdc.noaa.gov/billions/events. The total insured and uninsured direct losses considered include "physical damage to residential, commercial and government/municipal buildings, material assets within a building, time element losses (i.e., time-cost for businesses and hotel-costs for loss of living quarters), vehicles, public and private infrastructure, and agricultural assets (e.g., buildings, machinery, livestock)." The reported loss assessments do not include "losses to natural capital/assets, healthcare related losses, or values associated with loss of life" (Smith and Katz, 2013).

TABLE 1-1 Damage, Percent Damage, Frequency, and Percent Frequency by Disaster Type, 1980-2012, for All Billion-Dollar Weather- and Climate-Related Events in the United States (adjusted for inflation to 2013 dollars)

	No. of Disaster Events	No. of Deaths	Adjusted Damage (billion $)	% Damage	% Occurrence
Tropical cyclones	33	3,159	491.9	48.8	21.9
Droughts/heat waves	18	18,744	243.3	24.1	11.9
Severe local storms	55	1,391	111.8	11.1	36.4
Nontropical floods	17	397	86.4	8.6	11.3
Winter storms	10	882	29.8	3.0	6.6
Wildfires	12	151	23.6	2.3	7.9
Freezes	6	1	20.8	2.1	4.0
TOTAL	151	24,725	1,007.6	100	100

NOTE: Damage cost totals do not include 2013 events, from which damage data are not yet available.
SOURCE: Data from http://www.ncdc.noaa.gov/billions/events.

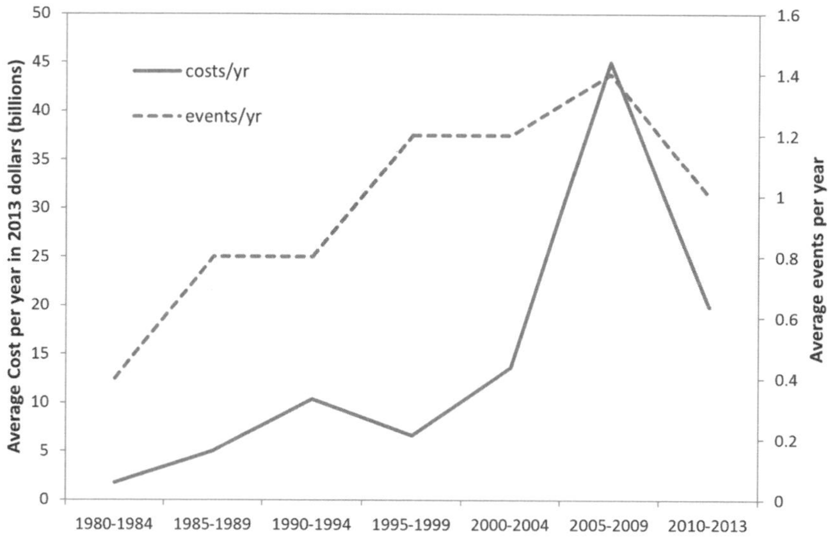

FIGURE 1-1 Average number and costs associated with billion-dollar coastal storm events in the United States between 1980 and 2013 by 5-year time increments.
SOURCE: Data from http://www.ncdc.noaa.gov/billions/events.

has grown rapidly for at least the past 80 years. Presently, coastal counties along the entire East and Gulf Coasts, which account for 24.2 percent of the U.S. population, grew by 11.4 percent from 2000 to 2012, essentially matching population growth across the United States. However, coastal population growth has been substantially skewed toward the southeast and Gulf of Mexico coastal areas, which grew by 20.8 and 17.8 percent, respectively, during this time period (Table 1-2). These areas are the most frequently impacted by hurricanes and tropical storms (Figure 1-2) and they typically have low topographic slope, meaning they lack the most effective natural defense against coastal storm surge and wave damage—vertical elevation of the land near the water's edge.

Natural cycles in tropical cyclone activity and observational bias in data sets before the modern satellite era (mid-1960s) make determining historical trends in either the frequency or severity of tropical cyclones during the past 100-150 years difficult (Landsea, 2007). However, despite clear increases in global mean temperature and tropical Atlantic sea surface temperature, statistically significant trends do not appear to exist in the number of Atlantic basin hurricanes or U.S. land-falling hurricanes since at least 1875 (Vecchi and Knutson, 2011; GFDL, 2013). Conversely, using six high-quality tide-gauge records (dating to 1923) from the southeastern United States, Grinsted et al. (2013) found that storm surge statistics were correlated with global temperature. This study identified a doubling of the likelihood of a Katrina-magnitude storm surge during the 20th century, which could be a significant finding because the oceanic response, represented by storm surge and waves, is usually the most destructive aspect of a hurricane.

TABLE 1-2 Population Growth in U.S. East Coast and Gulf of Mexico Coastal Counties

	Coastal County Population		
	2000	2012	% Change
Northeast	9,300,000	9,700,000	4.6
Mid Atlantic	32,200,000	34,400,000	6.9
Southeast	11,800,000	14,300,000	20.8
Gulf of Mexico	14,800,000	17,500,000	17.8
Total 4 regions	68,200,000	75,900,000	11.4
Total U.S.	281,400,000	313,900,000	11.5
East and Gulf Coast County Population as Percentage of Total U.S. Population	24.2	24.2	

SOURCE: Data from the National Ocean Economics Program, www.oceaneconomics.org.

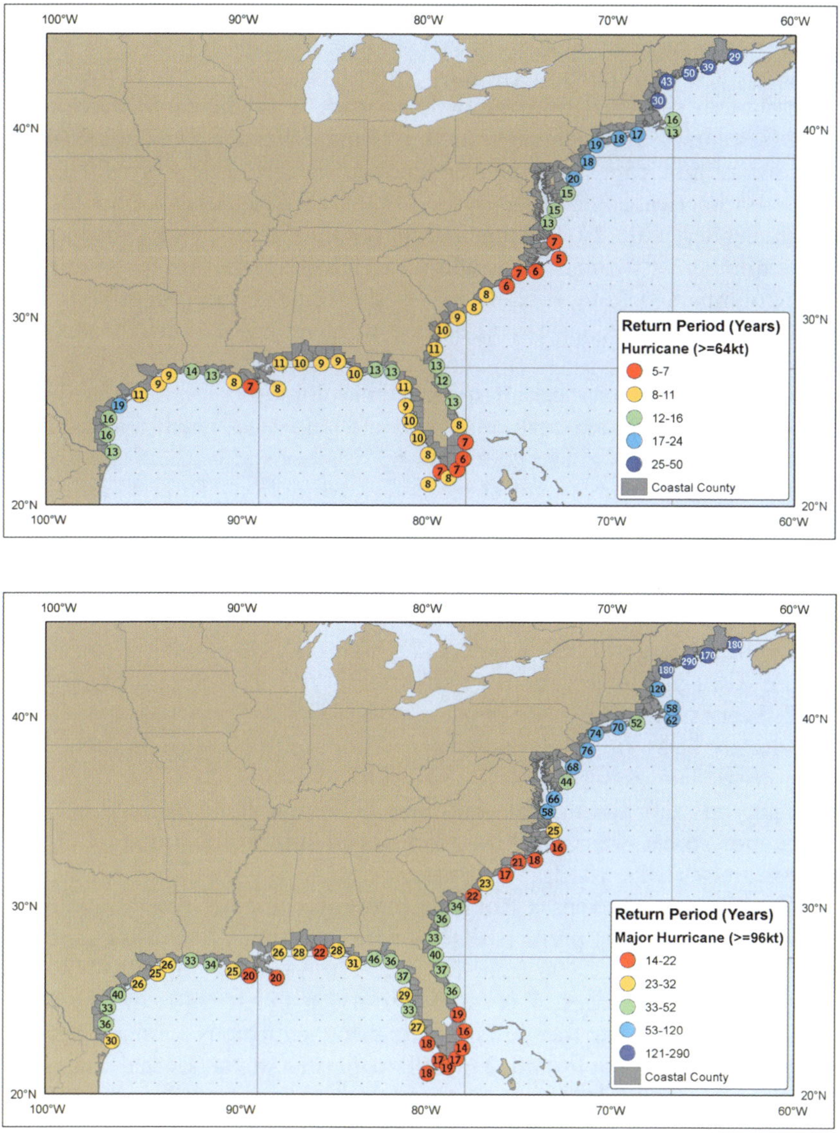

FIGURE 1-2 Estimated average return periods (years) for all hurricanes (top) and major hurricanes (category 3 or greater, bottom) passing within 50 nautical miles (90 km) of various locations on the U.S. East and Gulf of Mexico Coasts. SOURCE: Data from http://www.nhc.noaa.gov/climo/.

Although increases in coastal development in high hurricane hazard areas appear to have dominated the growth in coastal natural disaster–related economic losses for much of the past century, this may change in the future. Even though the total number of hurricanes is predicted to decrease in the 21st century, research suggests that climate warming may increase the intensity of hurricanes and the frequency of the strongest storms (i.e., category 4-5 hurricanes) (Bender et al., 2010; Emanuel, 2013; Knutson et al., 2013). Bender et al. (2010) estimated that in the Atlantic basin, the increase in the number of strong storms will outweigh the reduction in overall hurricane numbers yielding roughly a 30 percent increase in potential damage by 2100. Hurricanes are also projected to have higher rainfall rates than today's hurricanes.[3]

In addition to changes in storm climatology, sea-level rise is raising the level of the coastal ocean relative to the land. As a result, coastal cities are increasingly exposed to flooding, and beaches and wetlands are subject to deterioration from storm surge and wave action (Titus et al., 2009; Sallenger et al., 2012). Globally, relative sea-level rise—the observed change of sea level relative to land surface at a particular point, thereby considering other factors such as subsidence—has averaged 0.12 in/yr (3.1 mm/yr) over the past two decades. However, local rates of sea-level rise vary considerably, with the largest rates in the United States in areas of the northern and western Gulf of Mexico and the mid-Atlantic (Figure 1-3) (Sallenger et al., 2012; Ezer et al., 2013). The impacts of sea-level rise over the past century can already be seen in the frequency of flooding that occurs in many low-lying areas. For example, parts of Norfolk, Virginia, that saw significant flooding only during hurricanes in the 1930s, now flood during high tides and minor storm events and therefore spend substantial amounts of time underwater (VIMS, 2013).

The most recent report from the Intergovernmental Panel on Climate Change (IPCC, 2013) predicts that climate warming will cause a mean increase in sea level by 2100 from 1.4 to 2.4 ft (44 to 74 cm). NRC (2012c) predicted an even larger increase (1.7 to 4.6 ft [51 to 140 cm]) by 2100. These increases have the potential to bring enormous damage because nearly 5 million people and 2.6 million homes in the United States are found at less than 4 ft (1.2 m) above high tide (Climate Central, 2012).

Assuming that sea level rises by only 1.6 ft (0.5 m) by the year 2100, Sweet et al. (2013) calculated that the return periods for Hurricane-Sandy–level storm surges would be reduced by a factor of approximately 4, with higher sea-level rises further reducing the intervals between major inundation events. Lin et al. (2012) considered an ensemble of tropical storm

---

[3]See http://www.gfdl.noaa.gov/global-warming-and-hurricanes.

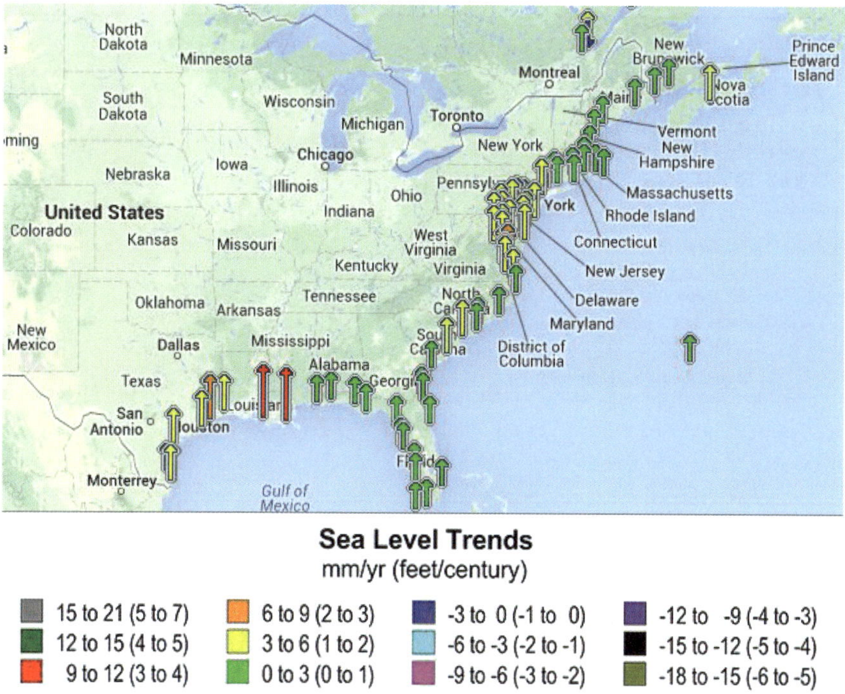

FIGURE 1-3 Rates of relative sea-level rise (mm/yr [ft/century]) along the U.S. East and Gulf Coasts.
SOURCE: Data from http://tidesandcurrents.noaa.gov/sltrends/sltrends.shtml.

scenarios associated with climate change and found that by 2100, assuming 3.3-ft (1-m) sea-level rise, today's 1 percent annual-chance (100-year) flooding event in the greater New York City area may increase in annual probability to 5 percent (a 20-year flood) or more. The simple addition of elevated sea levels and existing storm surge risk can be used to create approximate estimates of the spatial extent of areas that are at risk under future sea levels (Figure 1-4).

The recently released U.S. National Climate Assessment (Melillo et al., 2014) identifies the U.S. Gulf of Mexico and Atlantic Coasts as being subject to increased risk of storm surge damage and flooding due to sea-level rise combined with coastal storms. Impacts will occur to homes, critical infrastructure, cultural and historic resources, agriculture, ports, tourism, coastal resources, and coastal ecosystems. Vulnerability to these impacts is uneven due to socioeconomic disparities throughout the region.

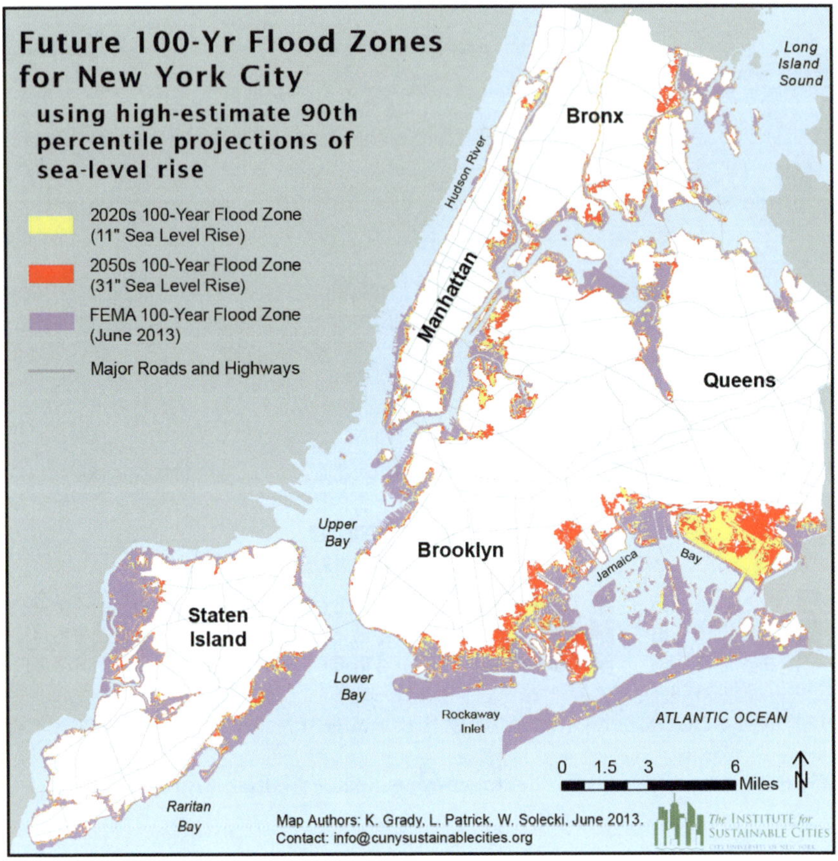

FIGURE 1-4 Current Federal Emergency Management Agency 1 percent chance (100-year) floodplain (purple) and an approximation of the extent of flooding with the same 1 percent probability considering two sea-level rise scenarios: 11 inches (28 cm; yellow) and 31 inches (79 cm; red).
SOURCE: Adapted, with permission, from NPCC (2013).

Shifting Federal Roles

Concurrent with the growth in natural hazard economic losses, there has also been a substantial shift in the source of funds used to cover these losses in the United States. The federal government's assistance to disaster victims is well illustrated by the large increase in the past 60 years in the number of Presidential disaster declarations that have occurred (from approximately 10 to nearly 100 per year for all weather-related disasters) and a similar relative increase (from approximately 1 to 10 per year) in coastal storm-related Presidential disaster declarations[4] (Figure 1-5). There has also been a substantial increase in the percentage of severe storm-related damages covered by federal aid over this period, from 6 percent for Hurricane Diane in 1955 to more than 75 percent for Hurricane Sandy (Table 1-3). Abundant federal assistance has raised concerns of a "moral hazard" in which state and local government leaders are discouraged from investing in disaster mitigation and preparedness because they expect to be "bailed out" by the federal government (Sylves and Buzas, 2007). Federal programs supporting coastal risk management and disaster recovery are discussed in Chapter 2.

Together, the growth in coastal disaster losses associated with population redistribution, the looming implications of climate change, including sea-level rise, and the shift in the fiscal responsibility for disasters illuminate pressing challenges ahead in coastal risk management.

RESILIENCE AND RISK

Full protection from coastal hazards and related damages is typically impractical at community to national scales. Even the largest levees or surge barriers could be overtopped by a large storm or suffer from structural failures. Thus, local, state, and federal governments are increasingly recognizing the importance of becoming more *resilient* to hazards and disasters, including coastal hazards. NRC (2012b) defines *resilience* as "the ability to prepare and plan for, absorb, recover from, or more successfully adapt to actual or potential adverse events." Resilience depends on the reliability of community service systems in the face of significant disturbance or the capability to recover those services within an acceptable time period, thereby enabling a community to maintain its economic, communications, transportation, social, political, and quality-of-life functions (Tierney et al., 2001; DHS, 2007; Tierney and Bruneau, 2007). Resilience planning, therefore, focuses on the specific needs of the community served and the capacity to provide the necessary services throughout the recovery period (Corotis, 2011; NRC, 2011a).

Resilient communities are able to assess and manage risks, are generally well informed of threats, and are clear about the roles and responsibilities of individuals and organizations in the community with respect to risk (NRC, 2012b). Resilient communities take into account both pre-disaster mitigation measures and post-disaster recovery measures to determine an appropriate allocation of resources to improve resilience within budgetary

---

[4]See http://www.fema.gov/disasters/grid/year.

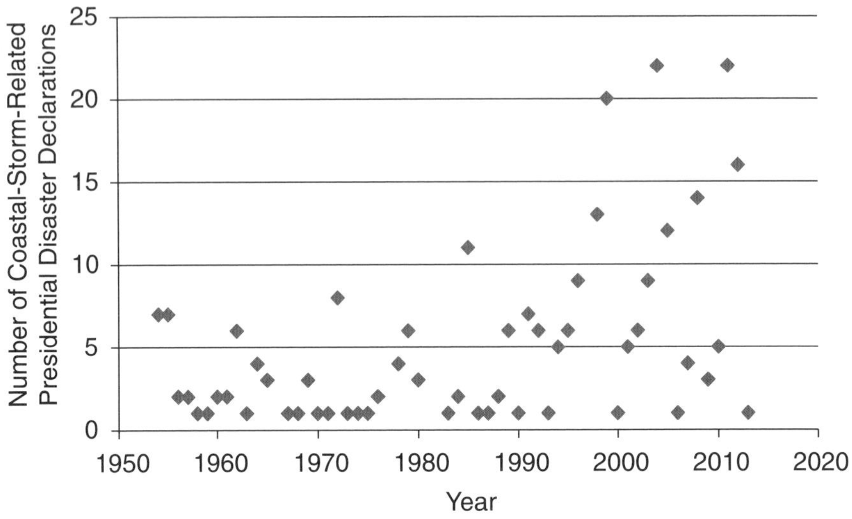

FIGURE 1-5 U.S. Presidential disaster declarations for hurricanes and coastal storms by year, 1953-2013.
SOURCE: Data from http://www.fema.gov/disasters/grid/year.

constraints. Pre-disaster mitigation can prevent property damage and some business and infrastructure impacts, but resilience can also be improved by strategies to recover more quickly (Rose et al., 2007). Actions to enhance resilience that can be implemented at the local level prior to a disaster include emergency planning drills and disaster planning for businesses (e.g., increasing inventories, identifying alternative supply-chain sources and operating locations). Other actions can be taken following a disaster, such as business relocation and conservation of critical supplies.

Understanding, managing, and reducing risk are foundations for building resilience. *Risk* is "the potential for hazards to cause adverse effects on

TABLE 1-3 Change in Percentage of Federal Aid Following Major Tropical Cyclones, from 1955-2012

Disaster	Federal Aid as a Percentage of Total Damage
Hurricane Sandy (2012)	>75
Hurricane Ike (2008)	69
Hurricane Katrina (2005)	50
Hurricane Hugo (1989)	23
Hurricane Diane (1955)	6

SOURCE: Michel-Kerjan (2013).

our lives; health; economic well-being; social, environmental, and cultural assets; infrastructure; and the services expected from institutions and the environment" (NRC, 2012b). In natural hazard and disaster fields, risk for a particular hazard, place, and time period is represented as the probability of a hazardous event multiplied by its consequence (Box 1-1; BSI, 2002; Gouldby and Samuels, 2005; UNISDR, 2009). *Hazard* refers to the physical event with the potential to result in harm (Gouldby and Samuels, 2005). Thus, flooding or overland waves caused by hurricanes or other strong coastal storms are the primary hazard—not the storm or the coastal storm surge itself. *Consequence* represents the impact caused by the hazard. Consequence can encompass a range of values, such as economic damage (monetary), number of people or properties affected, harm to individuals (e.g., fatalities, injuries, stress), and environmental impacts. Consequence is controlled by *exposure* (density of people, property, or other elements in hazard zones [UNISDR, 2009]) and *vulnerability* (a system's potential to be harmed, which is a function of both the susceptibility to experience harm and the value, expressed in monetary or other terms of the people,

BOX 1-1
Components of Risk

For purposes of quantitative risk assessment, risk is represented as the probability of a hazard multiplied by the consequence:

$$R = H \times C$$

in which R = *risk*, H = probability of the occurrence of a *hazard* (e.g., storm-induced flooding), and C = *consequence*. *Consequence* represents the impact and can be measured in various units including monetary damage, number of people or properties affected, harm to individuals (e.g., fatalities, injuries, stress), and environmental impacts. Consequence may be expressed as a function of the *exposure* (E) (the density of people, property, systems, or other elements present in hazard zones) and the *vulnerability* (V), which is a system's potential to be harmed:

$$C = f(E, V).$$

Vulnerability can be defined in terms of the susceptibility to harm and the value (in monetary or other terms) of the people, property, systems, or elements present in hazard zones.

SOURCE: Data from Gouldby and Samuels (2005).

property, or other elements in the hazard zones [Box 1-1; Gouldby and Samuels, 2005]).

Coastal risk reduction focusing on the hazard is typically achieved through hard structural measures (such as construction of seawalls or levees) or nature-based approaches (such as building dunes) to reduce the wave and flood hazard probability. Risk reduction focusing on the consequence is typically achieved by an array of measures that change exposure (e.g., relocating homes and businesses away from high-hazard areas or evacuating prior to a storm event) or reduce vulnerability (e.g., elevating structures or enhancing risk awareness) (see Table 1-4). In the past decades, much more attention has been placed on strategies that reduce the probability of flooding than those that reduce exposure to storm events (i.e., the extent to which we live in harm's way) (NRC, 2012b). To improve coastal risk management, it will be important to consider options that will address both sides of this risk-exposure equation.

Risk management is a continuous process that identifies the hazard(s) facing a community, assesses the risk from these hazards (Box 1-2), develops and implements risk reduction (mitigation) measures, reevaluates and reviews these measures, and develops and adjusts risk policies. If done well, risk management should help build capacity for communities to become more resilient to disasters (NRC, 2012b). For example, risk reduction efforts that place more value on critical infrastructure (e.g., hospitals, water and wastewater treatment facilities, power plants) than on other infrastruc-

TABLE 1-4 Risk Reduction Measures Linked to Components of Risk Reduction

	Risk Reduction		
	Probability of Hazard (Flooding, Wave Damage)	Consequence	
Coastal Risk Mitigation Measures		Exposure	Vulnerability
Surge barriers	X		
Levees, sea walls	X		
Beach nourishment and dune building	X		
Relocation		X	
Land-use restrictions		X	
Elevating and flood-proofing structures			X
Flood warning and preparedness programs		X	X
Flood insurance[a]			X

[a]If flood insurance is appropriately priced, the result should communicate risk and may spur additional mitigation measures, thus reducing vulnerability in addition to transferring risk to a broader risk pool.

BOX 1-2
Evolution of Coastal Risk Assessment

Early coastal risk assessment in the United States was based on deterministic characterizations of hazards, invoking "design storms" (e.g., the *standard project hurricane*—the most severe storm reasonably characteristic of the project area—or the *probable maximum hurricane*—the most severe storm thought possible in the project area [Graham and Nunn, 1959; NOAA, 1979; Woolley and Shabman, 2008]) that were presumed to be appropriate cases for design. Within the United States, the U.S. Army Corps of Engineers (USACE) primarily used the Standard Project Hurricane to set design water levels and associated hazards in its design projects, while the Nuclear Regulatory Commission used the probable maximum hurricane for this purpose. The stipulation that the USACE design event was linked to a storm that was only "reasonably characteristic of the project" area, rather than the maximum possible storm, implied a recognition that such designs were potentially vulnerable to future storms; however, risk assessment approaches in that era considered only a few discrete failure modes and their outcomes. Although this design simplification was consistent with the approach followed in other large engineering projects at that time, it ignored the full range of storm characteristics, uncertainty in the performance of levees and other protective systems under storm loading, and the likelihood of a particular hazard result. Because of the shortcomings of deterministic risk assessment methodologies, *probabilistic risk assessment* has become the basis of modern risk assessment. Probabilistic risk assessment uses quantitative calculations and models to compute the probabilities that certain hazards occur, the systems' response to those hazards, and the consequences associated with adverse outcomes of the systems' response. Thus, its results show not only what could happen, but also how likely each outcome is to occur. Under good professional practice, uncertainty is also quantified and integrated into the decision process (NRC, 1994; IOM, 2013).

ture can help improve community resilience by allowing more rapid recovery from a disaster with less disruption to critical services. The impacts of Hurricane Sandy have led to recommendations for increased consideration of critical infrastructure in comprehensive coastal risk assessments and risk reduction planning (USACE, 2013d).

Even after risk reduction measures are taken, some risk will remain because no risk reduction measure ever provides absolute protection. The risk that remains is referred to as *residual risk*. In the coastal zone, residual risk exists because storms larger than those designed for may occur, or the risk reduction measures put in place have a possibility of failing to perform as designed. Communities can work collectively to determine an acceptable level of residual risk based on their risk tolerance and the benefits and costs of additional risk reduction measures (see Chapter 4).

Once calculated, risk can be used in several different decision-making approaches. A risk-standard (or "level-of-protection") approach recommends investment in coastal risk reduction measures to drive residual risk below a specified level (such as a 1 percent annual chance of exceedance, also known as a 100-year event; Box 1-3). Congress specified the use of a 1 percent annual chance of exceedance as the design basis of the Hurricane and Storm Damage Risk Reduction System around greater New Orleans (USACE, 2013b). A benefit-cost approach determines worthy coastal risk management investments based on a comparison of benefits (measured as the value of risk reduction) to the costs of investment. Hybrid approaches are also common (see Chapter 4).

Application of either approach requires careful consideration of the long-term effects of risk reduction strategies on overall risk. Measures designed to reduce risk by decreasing the probability of the hazard, may encourage increased exposure (e.g., additional development or redevelopment) and/or increased vulnerability (e.g., higher-priced homes, risk complacency) in the hazard area and, in the long run, lead to higher risk. These risk reduction measures may thus decrease the negative consequences of small or moderate events, but increase the negative consequences of catastrophic events (Box 1-4; Hallegatte, 2012; NRC, 2013). Elevating homes in a coastal area above storm surge levels may reduce vulnerability but encourage expanded development, thereby increasing exposure to severe floods as well as other hazards such as wind or coastal erosion. Also, the expanded development may encourage investment in public infrastructure (e.g., roads, water, sewer, communications, emergency services) that are then subject to hazard damage.

COASTAL RISK MANAGEMENT STRATEGIES AND MEASURES

Numerous designs and strategies can be used to mitigate coastal risk associated with severe storms. These include measures to reduce the hazard, such as seawalls, breakwaters, and levees; natural and nature-based features, including wetlands, natural and replenished dunes, and mangrove forests; and strategies to reduce the consequences of an event, such as land-use planning, floodproofing, and relocation (USACE, 2013a).

The primary hazards under consideration in this report are flooding and wave attack. Mitigation of coastal flooding during severe storms is largely dependent on defending against or reducing the vulnerability to storm surge. Many oceanic responses, including wind waves, swell, tides, and surge, can be classified under the general category of waves. However, throughout this report (predominantly below and in Chapter 3) *waves* are considered to represent only relatively short time- and spatial-scale responses to wind forcing that pass a given location in a matter of seconds

BOX 1-3
History of the 1 Percent Annual-Chance (100-Year) Flood

The concept of the "100-year" event is ever present in the probabilistic characterization of natural hazards. In recent years the trend has been to call this the "1 percent chance" event, to emphasize that the event could happen at any time.

Although considerations of annual flood hazard criteria arose in the United States by the mid-20th century (ASFPM Foundation, 2004), Executive Order 11296, signed by President Johnson in 1966, first directed federal agencies to take flood probabilities into account when making decisions in locating federally owned buildings and roads. Shortly after, in 1968, the National Flood Insurance Program (NFIP) was established to reduce future flood damages and federal disaster assistance expenditures through community-based floodplain ordinances and flood insurance (NRC, 2013). Neither the executive order nor the National Flood Insurance Act of 1968, however, defined a standard criterion for flood hazard areas.

In December 1968, a special committee of experts convened by the University of Chicago recommended that the 1 percent chance (100-year) event be considered an initial standard for the NFIP, and the Flood Insurance Administration formally established the 1 percent chance event as the regulatory standard for the NFIP in 1971 (Wright, 2000; Galloway et al., 2006). Purchase of flood insurance was required to obtain a mortgage from a federally regulated or insured lender for those living in the 1 percent chance (100-year) flood hazard area, although this requirement was later waived for properties located behind structures designed to protect against such an event. In 1972, the Federal Water Resources Council recommended that agencies use the 1 percent chance event as the baseline flood in floodplain usage decisions, although other standards were permitted when appropriate (Robinson, 2004). The 1 percent chance event was selected "because it was already being used by some agencies, and because it was thought that a flood of that magnitude and frequency represented both a reasonable probability of occurrence, a loss worth protecting against and an intermediate level that would alert planners and property owners to the effects of even greater floods" (Robinson, 2004). It did not represent an attempt to achieve optimal balancing of risks and benefits. Ultimately, it represented a compromise between decision makers and those who would be affected by its implementation, and it provided "a point of departure for adjustments that could reflect the differences that might exist in floodplains across the country and in the objectives of the States and localities that would implement the standard" (Galloway et al., 2006).

Coastal flood standards in many developed countries are far stronger (less probable) than the 1 percent chance event. For example the Netherlands and Japan use the 0.01 percent chance (10,000-year) event for some coastal works (Galloway et al., 2006), although the derivation of the flood level for such a rare event from limited duration observations introduces inherent uncertainties. Methods for dealing with those uncertainties have been (Roscoe and Diermanse, 2011) and continue to be developed. Many U.S. studies have concluded that the 1 percent chance event is inadequate as a flood risk reduction design basis for urban areas (e.g., Galloway et al., 2006; ASFPM, 2007; NRC 2009).

BOX 1-4
New Orleans:
Flood Protection Led to Increased At-Risk Development

In 1965, after flood damages from Hurricane Betsy, Congress authorized a hurricane protection levee project along Lake Pontchartrain and vicinity, designed to protect the main urban areas of New Orleans from flooding from a Standard Project Hurricane that was described as having a likelihood of occurrence of approximately 1 in 200 years and the characteristics of a fast-moving category 3 hurricane (USACE, 1965; GAO, 2005). A feasibility study of the project found that flood protection for existing development accounted for 21 percent of the benefits, while the remaining 79 percent was associated with flood protection for new development, made possible by the enhanced levee system (GAO, 1976). In the decade after authorization of the Lake Pontchartrain project, Jefferson Parish added 47,000 housing units and Orleans Parish added 29,000 in the former low-lying wetland areas.

The development of the area east of the Industrial Canal, which contains 50 percent of New Orleans' land area, is especially suggestive of the interaction between flood risk reduction measures and development. In 1960, before the new levee plan, eastern New Orleans consisted of a few scattered residential and commercial structures. In anticipation of construction of Interstate 10 and the extension of the city's levee system, the city adopted a comprehensive plan in 1966 that designated the area for intensive urban development. In the 1970s, this area experienced development of 22,000 new housing units. In a retrospective assessment of the area's development trends, the city's 1999 land-use plan stated: "Full scale development ensued . . . (and) the area continued to grow from 1975 to 1985. New subdivisions were developed at a rapid pace . . . (and) major commercial centers developed and prospered." (New Orleans City Planning Commission, 1999). In 2005, the entire area of urban growth that was proposed to be reasonably safe because of levee investments was flooded by Hurricane Katrina due to overtopping of design levels and structural failures at levels below the project design (IPET, 2009; Figure 1-4-1). Altogether, Hurricane Katrina caused over $148 billion in damages (in 2013 dollars) and 1,833 deaths.[a]

to minutes and have wave lengths measured in feet (or meters). These are commonly called wind waves or swell. *Storm surge* represents a much larger- and longer-scale response, sometimes inundating an area for hours, with wave lengths measured in miles (or kilometers). Storm surge is caused by the combination of winds, atmospheric pressure, the rotation of the earth, and wave-induced setup (Dean and Dalrymple, 2002). At any given time, the total coastal water level is composed of the astronomical tide plus storm surge, wave height, and freshwater input (if important).[5] Due to their

[5]See http://www.nws.noaa.gov/om/hurricane/resources/surge_intro.pdf.

FIGURE 1-4-1 Flooding of New Orleans after Hurricane Katrina.

[a]See http://www.ncdc.noaa.gov/billions/events.
SOURCE: NOAA. Available at http://www.mississippiriverdelta.org/files/2012/03/800px-EasternNewOrleansFlood11Sept2005NOAA.jpg.

very different time and spatial scales, storm surge and waves respond quite differently to hazard mitigation strategies and therefore these responses are discussed individually.

Measures to Reduce the Hazard—Hard Structures

Hard structural measures to address coastal storm hazards are typically static, engineered features designed to reduce wave damage and flooding, and they may also decrease shoreline erosion. Sometimes termed "gray

FIGURE 1-6 Seawall along the coast of Galveston Island, Texas.
SOURCE: Photo courtesy of Melanie Fitzpatrick, Union of Concerned Scientists.

infrastructure" or "hard engineering," these structures include seawalls, levees and floodwalls, and surge barriers:

- **Seawalls** are constructed parallel to the shoreline to reduce impacts from storm surge and waves to developed lands behind the seawall. Seawalls may be vertical or curved walls (Figure 1-6) or designed as a mound built from rock or concrete blocks. The seawall reflects wave energy back to the sea, and therefore can increase erosion on the coastal side of the wall. Depending on lateral currents, seawalls may also cause increased erosion of adjacent, unprotected coastal areas.
- **Levees and floodwalls** are onshore engineered structures most commonly constructed along riverine floodplains that are designed "to contain, control, or divert the flow of water so as to provide protection from temporary flooding" (44 CFR § 59.1). Levees (sometimes also called dikes) are typically wide earthen embankments that are designed to control flooding over a large area up to a specific water level. Levees, however, can also be used in coastal settings, where

they may be paired with other mitigation features, such as revetments or coastal wetlands that buffer the levee against erosive wave forces. Floodwalls—typically vertical concrete walls—are usually constructed in areas where there is insufficient space for the wide footprint of an earthen levee. Floodwalls can also be constructed on top of a levee when space limits any further expansion of the levee footprint that would be required for increasing the height of the levee itself (Figure 1-7).

- **Storm surge barriers** are designed to block storm surges from propagating inland via rivers or other waterways (Figure 1-8). Gates in the barriers are left open to allow water to flow through under normal conditions but can be closed when storm surges are expected.

Other engineered measures, such as breakwaters (offshore rock mounds or concrete armor units), revetments (onshore armoring constructed of stone or concrete), and bulkheads (short vertical walls common in estuarine settings) are primarily intended to reduce coastal erosion but also serve to reduce wave energy that accompanies storm surge. Hard structural coastal risk reduction measures were commonly used by the USACE in the 1950s and 1960s, but their use decreased beginning in the 1970s (Figure 1-9).

Measures to Reduce the Hazard—Natural and Nature-Based Features

The presence of natural features, such as barrier islands, vegetated dunes, coastal wetlands, mangrove forests, and reefs, may reduce coastal storm hazards by attenuating wave energy and storm surge and possibly stabilizing sediment. However, the effectiveness of these features depends on the specific characteristics of the storm and the features themselves. Dunes serve as a physical barrier blocking storm surge, although their longevity depends on the adjacent beach slope, the sediment characteristics, the height and width of the dune, and the extent of dune vegetation. Coastal wetland vegetation and the land it helps retain may reduce the rate of storm surge advancement and extent (Wamsley et al., 2010). Mangroves are capable of damping incident waves, reducing wind speed within the canopy, and potentially reducing storm surge, depending on their lateral extent (see Chapter 3; McIvor et al., 2012; Zhang et al., 2012).

Nature-based coastal risk reduction strategies are designed and engineered to mimic natural features for the purpose of attenuating storm surge. The most commonly applied coastal risk reduction strategies in the United States are dune building and beach nourishment (also called beach fill), (Figure 1-10). Periodic nourishment with sand from offshore locations creates a wide beach area to absorb the energy of breaking waves, and replenished sand dunes can serve as a physical barrier, albeit a dynamic one,

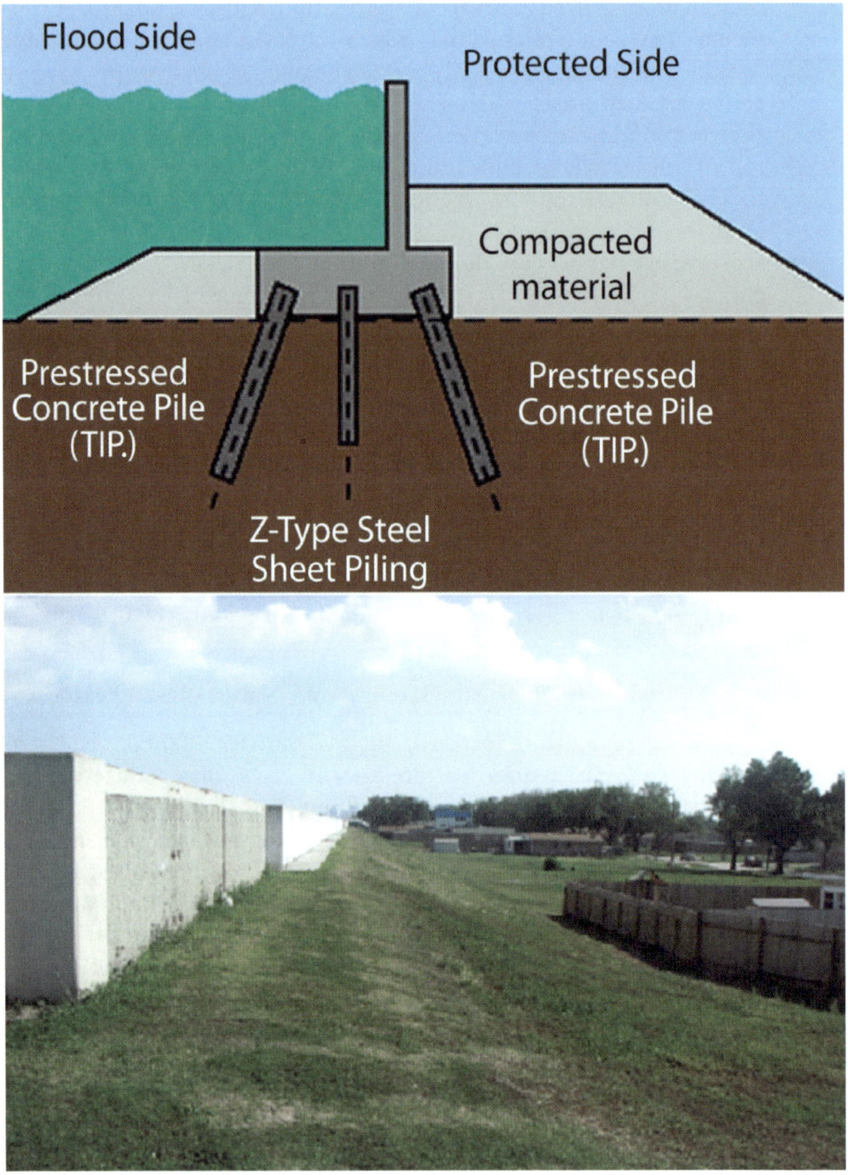

FIGURE 1-7 Schematic of a floodwall paired with a levee for flood risk reduction along a canal (top) and a levee with floodwall along the London Avenue Canal in New Orleans, Louisiana (bottom).
SOURCES: USACE (http://library.water-resources.us/docs/MMDL/FLD/Feature.cfm?ID=2);http://en.wikipedia.org/wiki/File:London_West_from_Robert_E_Lee_to_DPS4_0001.jpg).

FIGURE 1-8 Fox Point hurricane barrier on the Providence River in Providence, Rhode Island, built in 1966.
SOURCE: Photo courtesy of Neil Aquino.

to reduce flooding and destructive wave energy on structures located behind the dunes. However, replenishment brings additional ecological impacts, as discussed in Chapter 3, and replenished beaches may have a different slope than natural beaches, which can alter the incident wave conditions. Other potential nature-based coastal risk reduction strategies include conservation and/or construction of wetlands and oyster reefs, which may also provide additional ecosystem services benefits.

Measures to Reduce the Consequences

Consequence reduction measures aim to reduce the exposure or vulnerability to a hazard. These approaches include elevating and floodproofing structures (and related building codes) and nonstructural strategies, such as flood warning and emergency preparedness programs, flood insurance, land-use regulations, restrictions on development in areas of severe flood hazard, and property acquisition and relocation programs (see Table 1-4). These strategies are sometimes broadly called "nonstructural" measures,

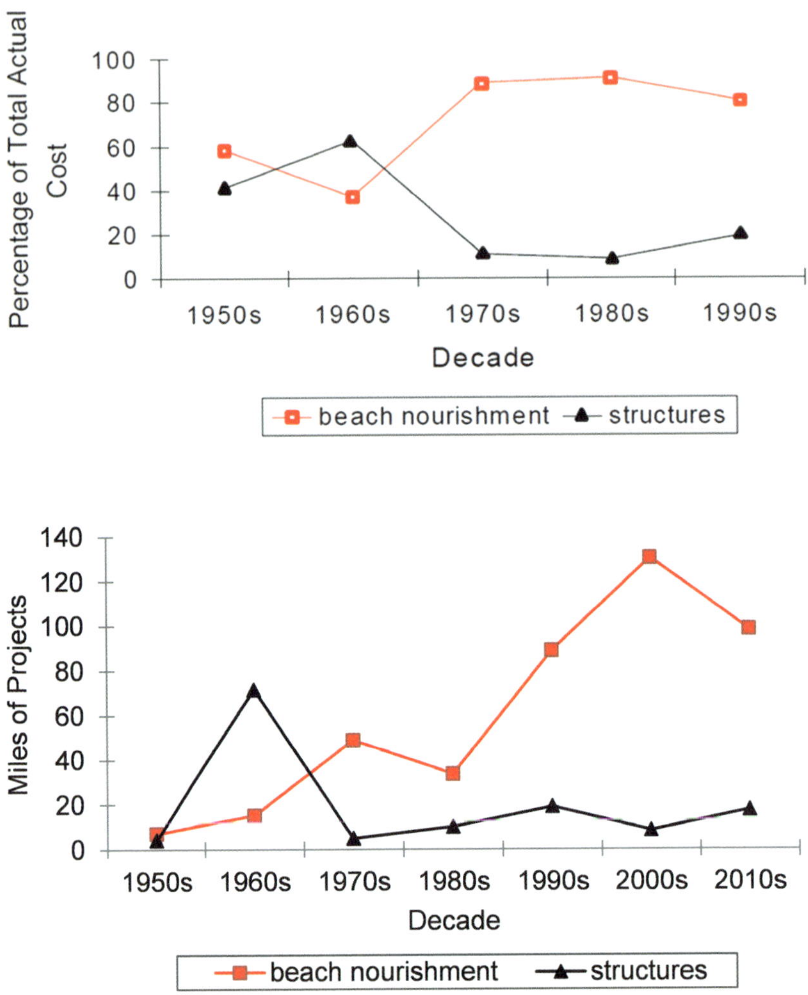

FIGURE 1-9 Percentage of total USACE coastal risk reduction expenditures (top) between hard structural measures and beach nourishment (including dune building) projects and miles of project (bottom) by decade. Recent cost data are not available, but the percentage of overall coastal risk reduction costs represented by hard structural measures has likely increased in the past decade with the post-Katrina construction of the Hurricane Storm Damage Risk Reduction System. Data used to compile the bottom figure are listed in Appendix B.
SOURCE: USACE (1996); D. Cresitello, USACE, personal communication, 2014.

FIGURE 1-10 Beach nourishment in Ocean City, New Jersey.
SOURCE: NOAA (2007).

although to minimize confusion, for the purposes of this report, the term "nonstructural" does not include floodproofing and elevation of individual structures.

Flood preparedness programs might include delineation of flood hazard areas, effective communication of risks to community residents and developers, development and communication of evacuation plans, and flood insurance for those at risk of flooding. If appropriately priced, flood insurance serves as both a risk transfer mechanism and an effective risk communication tool. Additionally, detailed and accurate forecasts and flood warning systems are essential for officials and citizens to be able to plan for and respond to a flood event, including decisions regarding evacuation (NRC, 2012b).

Flood-related impacts can also be minimized through well-enforced building codes and land-use regulations. Communities can restrict development in severe flood hazard areas and limit the construction of new public infrastructure that facilitates development (e.g., utilities, transportation). Additionally, communities can develop plans for relocating existing critical infrastructure to less risky locations, either when aging facilities require replacement or when facilities are severely damaged by coastal storms. Local governments can require elevation and other floodproofing measures in all new construction, although new building codes will take time to produce widespread changes and the codes must be enforced if they are to be effective. Kunreuther (1996) found that one-third of the damage from Hurricane Andrew could have been avoided if the state and local building codes had been enforced. Existing structures in floodprone areas can be elevated so that the main floor is above the base flood elevation (Figure 1-11) or residents with repeated flood damage can be encouraged through economic incentives to relocate.

STATEMENT OF TASK AND REPORT STRUCTURE

This study was undertaken as part of a broad 5-year effort to provide advice to the USACE on a range of scientific, engineering, and water resources planning issues through periodic reports. Prior to this current emphasis on coastal risk reduction, the NRC Committee on U.S. Army Corps of Engineers Water Resources Science, Engineering, and Planning issued two reports: *National Water Resources Challenges Facing the U.S. Army Corps of Engineers* (NRC, 2011b) and *Corps of Engineers Water Resources Infrastructure: Deterioration, Investment, or Divestment?* (NRC, 2012a). The committee was subsequently reconstituted for specific focus on reducing flood risks from coastal storm surges along the East and Gulf Coasts and was tasked to address the following questions:

1. What coastal risk-reduction strategies have been used along the U.S. East and Gulf Coasts to reduce impacts of coastal flooding associated with storm surges, and what design standards or levels of protection have been used? To what extent have these many strategies and levels of protection proven effective in terms of economic return, protection of life safety, and minimizing environmental effects?
2. What are the regional and national implications of expanding the extent and levels of coastal storm surge protection? Examples might include operations and maintenance costs, sediment availability, and regional-scale sediment dynamics.
3. How might risk-related principles contribute to the development of design standards for coastal risk reduction projects? How might risk-related principles increase the ability of coastal regions and communities to prepare for coastal storms and surge, and adjust to changing coastal dynamics, such as prospects of sea-level rise?
4. What general principles might be used to guide future investments in U.S. coastal risk reduction?

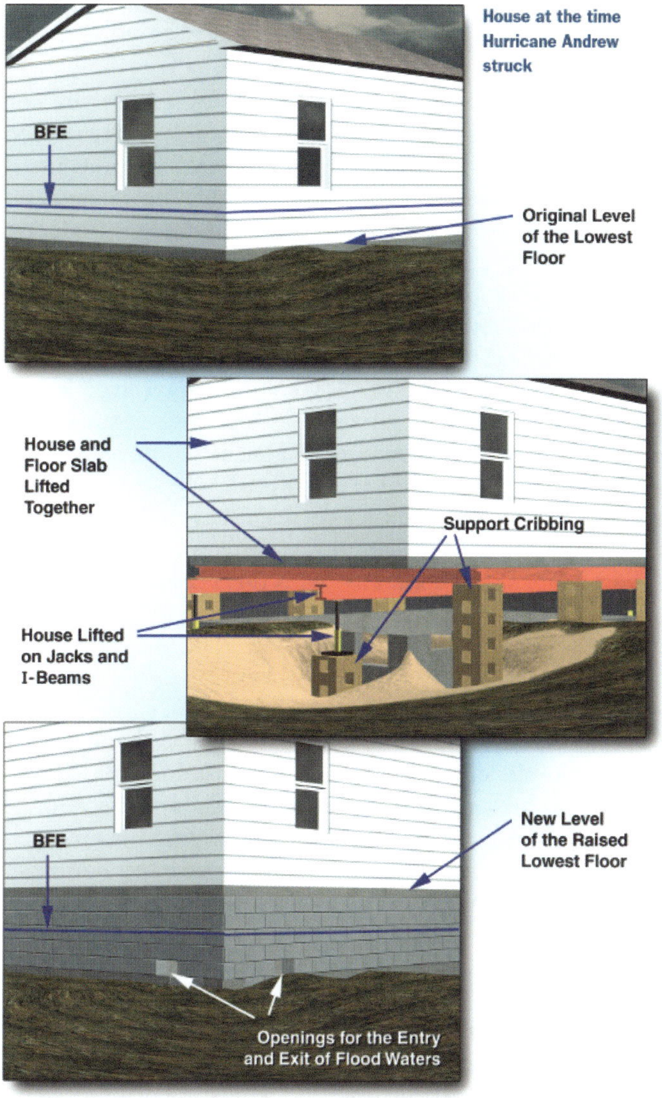

FIGURE 1-11 Example of one approach for elevating a house above the base flood elevation (BFE).
SOURCE: FEMA (2000).

The committee's charge specifically addresses coastal storms (hurricanes, tropical storms, and extratropical storms) and associated waves, storm surge, and flooding, which in the United States primarily affect the East and Gulf Coasts. Although Hawaii, Puerto Rico, and, under rare conditions, California are also subject to such storms, the study charge is focused on the East and Gulf Coasts. However, the committee's approach to Tasks #3 and 4 more broadly considers coastal storm surge risks throughout the nation. Other coastal hazards, such as erosion from mild or moderate storms, wind damage, or tsunami-induced flooding, were not considered in depth.

The committee's report and its conclusions and recommendations are based on a review of relevant technical literature, briefings, and discussions at its five meetings, and the experience and knowledge of the committee members in their fields of expertise. The committee received briefings from a range of federal and nonfederal agencies and organizations involved in coastal risk management (see Acknowledgments). However, because this study was conducted as part of a 5-year USACE-sponsored effort, the committee paid particular attention to the role of and opportunities for the USACE and, more broadly, the federal government in coastal risk reduction. The project scope combined with the 13-month study period did not allow the committee to give equal attention to all federal agencies or provide detailed discussion of actions that could be taken to reduce coastal risk at state or local levels.

In some cases the availability of data limited the extent to which these questions could be addressed. For example, the limited availability of retrospective analyses of the costs and benefits of coastal risk reduction projects after a storm event prevented a thorough analysis of the economic aspects of Task 1. The committee also found that Task 2 could not be answered quantitatively, because a full discussion of regional and national implications of expanding coastal risk reduction, particularly with respect to costs, would require detailed information on current risks and possible risk reduction strategies that was not available.

Following this introduction, the statement of task is addressed in three subsequent chapters of this report:

- Chapter 2 presents the institutional landscape for coastal risk management in the United States, highlighting major programs and recent budgets, and discusses the mechanisms by which the USACE develops and implements coastal risk reduction projects.
- Chapter 3 summarizes the current state of knowledge on the effectiveness of coastal risk reduction measures based on proven performance under coastal storms. The chapter includes discussion of financial and environmental benefits, costs, associated adverse

impacts, and regional implications for sediment availability (Tasks 1 and 2).

- Chapter 4 outlines key principles to guide future investments in coastal risk reduction, including a discussion of a benefit-cost approach constrained by acceptable risk for prioritizing coastal risk measures at a regional or a national scale (Tasks 3 and 4).
- Chapter 5 offers recommendations to enhance coastal storm risk management (Task 3).

2

Institutional Landscape for Coastal Risk Management

Responsibilities as they relate to coastal storm risks are shared among numerous institutions within local, state, and federal governments. Planning, zoning, and building ordinances—key elements of disaster preparedness—are primarily the responsibilities of local governments. Mitigation measures, such as raising homes and other coastal risk reduction strategies, can involve federal, state, and local agencies in varying capacities. Response and recovery following a major event involves numerous federal agencies to assist local and state governments. Additionally, several federal agencies provide data and tools to support planning at national, regional, and local levels. The private sector and nongovernmental organizations also have roles in risk management and disaster recovery, particularly at the community level. This chapter describes the major roles of federal agencies in coastal risk management, the roles and responsibilities typically borne by state and local governments, and federal actions that seek to provide a consistent national approach across these diverse programs. A detailed description of the U.S. Army Corps of Engineers (USACE) planning, authorization, and funding process is included to provide context for this complex landscape of coastal risk reduction efforts. The chapter concludes with a discussion of the alignment of current responsibilities, resources, risks, and rewards, with regard to coastal risk management.

FEDERAL AGENCY ROLES IN COASTAL RISK MANAGEMENT

Numerous federal agencies have roles in coastal risk management in the United States as reflected in legislation, executive action, and agency initiatives. There are four key agencies that share the bulk of the responsibility: the USACE, the Federal Emergency Management Agency (FEMA), the Department of Housing and Urban Development (HUD), and the National Oceanic and Atmospheric Administration (NOAA). Carter (2012) describes the USACE as the "principal federal agency involved in federal flood management investments and activities and flood-fighting." FEMA, under the Stafford Act and other legislation, has primary responsibility for disaster assistance and mitigation efforts and federally backed flood insurance. HUD provides funding for economic recovery of communities after a disaster, especially within low- and moderate-income populations. NOAA provides critical weather and climate information, as well as decision support tools to assist state and local coastal resource managers to assess potential impacts of storms.

A substantial amount of the funding for federal hazard and coastal risk–related programs is provided in response to national emergencies such as Hurricanes Katrina, Irene, and Sandy, rather than through annual appropriations. A summary of agencies and programs funded by the 2013 Hurricane Sandy Disaster Relief Act in Table 2-1 provides an illustrative snapshot of the number of agencies involved. It also reflects the significance of investments in federal housing programs, transportation, small business and public health programs, and other programs not traditionally associated with coastal hazard management and recovery.

This section outlines the major federal hazard management programs within the major agencies that contribute to various elements of coastal risk reduction and discusses recent federal coordination efforts.

U.S. Army Corps of Engineers

Hurricane and storm damage reduction is only one of the missions of the USACE that shapes the agency's coastal risk reduction efforts. Other related USACE missions include flood risk management, emergency operations, ecosystem restoration, and interagency and international services. On the basis of a range of authorities (see Box 2-1), the USACE—with congressional appropriation—works with local sponsors to examine the feasibility of coastal risk reduction–related projects, ranging from beach nourishment to barrier island restoration to engineered storm barriers (see Chapter 1). The USACE designs and constructs these projects contingent upon project-specific congressional authorization and appropriations. A list of USACE coastal storm risk management projects on the East and Gulf Coasts is provided in Appendix B. The USACE has also been tasked to undertake coastal risk reduction studies or efforts under specific authorizations limited to a particular area or event.

TABLE 2-1 Federal Agencies and Programs Funded by the 2013 Hurricane Sandy Disaster Relief Act

Agency	Component	Program or Appropriation Account	Amount (million $)[a]
Department of Housing and Urban Development (HUD)	Community Planning and Development	Community Development Fund	16,000
Department of Transportation (DOT)	Federal Transit Administration	Public Transportation Emergency Relief Program	10,900
	Federal Highway Administration	Federal-Aid Highways—Emergency Relief Program	2,022
	Federal Railroad Administration, Federal Aviation Administration		148
Department of Homeland Security (DHS)	Federal Emergency Management Agency	Disaster Relief Fund	11,488
		Disaster Assistance Direct Loan Program Account	300
	U.S. Coast Guard	Acquisitions, Construction, and Improvements	274
Department of the Army	USACE	Construction, flood control and coastal emergencies, and other	5,350
Department of the Interior (DOI)	Office of the Secretary	Departmental operations	360
	National Park Service	Construction and preservation	398
	Fish and Wildlife Service	Construction	68
Small Business Administration (SBA)		Disaster Loans Program Account, salaries, and expenses	804
Department of Health and Human Services (HHS)		Office of the Secretary, Public Health and Social Services Emergency Fund	800

continued

TABLE 2-1 Continued

Agency	Component	Program or Appropriation Account	Amount (million $)[a]
Environmental Protection Agency (EPA)		State and Tribal Assistance Grants	600
		Leaking Underground Storage Tank Fund	5
Department of Commerce (DOC)	National Oceanic and Atmospheric Administration	Construction; Operations, Research, and Facilities	326
Department of Veterans Affairs		Administration, construction, medical services	237
Department of Agriculture (USDA)		Emergency conservation, commodity assistance, capital improvement	228
Department of Defense	Navy, Army, Air Force, National Guard	Military construction, operation and maintenance, and management funds	113
Department of Labor	Employment and Training Administration	Training and Employment Services	25
Department of Justice	Federal Prison System and Federal Bureau of Investigation	Buildings, Facilities, Salaries, and Expenses	21
National Aeronautics and Space Administration (NASA)		Construction and Environmental Compliance and Restoration	15
General Services Administration (GSA)		Real Property Activities, Federal Buildings Fund	7
Total amount of appropriations			50,510

[a]The majority of appropriation accounts that received funding under the Disaster Relief Act were categorized as nondefense discretionary spending and therefore were subject to an additional reduction of 5.0 percent of their budgetary resources due to sequestration, not reflected here. Accounts that were categorized as nondefense mandatory spending were subject to a 5.1 percent reduction, and accounts that were categorized as defense discretionary spending were subject to a 7.8 percent reduction. Some accounts were exempt from sequestration as well. The actual sequestration of Disaster Relief Act funds in a program, project, or activity within an account may vary, depending on other sources of sequestrable funding in the program.

NOTE: Only those programs receiving $5 million or greater are included here. SOURCE: Modified from GAO (2013).

BOX 2-1
Evolution of USACE Authorities Related to Shore Protection and Coastal Risk Reduction

- Rivers and Harbors Act (1930)—Authorizes USACE to conduct shore erosion control studies.
- Flood Control Act (1946)—Authorizes emergency bank-protection works.
- P.L. 79-727 (1946)—Establishes federal policy to assist in the construction, but not maintenance, of works to protect publicly owned shores of the United States against erosion.
- P.L. 84-99 (1955)—Authorizes emergency management activities, including disaster preparedness, emergency response, and protection or repair of threatened shore protection works that are federally authorized.
- P.L. 84-826 (1956)—Provides federal assistance for periodic beach nourishment on the same basis as new construction, for a period to be specified by the Chief of Engineers.
- P.L. 86-645 (1960)—Authorizes the USACE to provide planning guidance and technical services to state, regional, and local governments at full federal expense to improve floodplain management.
- P.L. 87-874 (1962)—Increases the proportion of construction costs borne by the federal government for beach erosion control and shore protection projects.
- P.L. 89-72 (1965)—Specifies that recreation benefits shall be taken into account in determining the overall benefits.
- P.L. 90-483 (1968)—Section 111 authorizes to study, plan, and implement structural and nonstructural measures for the mitigation of shore damages attributable to federal navigation works.
- Water Resources Development Act (WRDA, 1974)—Section 22 authorizes the USACE to provide technical planning assistance (with 50 percent federal cost share) related to water resources development, including flood risk reduction.
- WRDA (1976)—Section 156 authorizes extension of federal participation in periodic beach nourishment, up to 15 years from initiation of construction.
- WRDA (1986)—Section 103(d) specifies cost sharing for various project purposes. Section 934 increases to 50 years the authorized period of time federal participation can be extended in periodic beach nourishment after the date of initiation of construction.
- WRDA (1999)—Section 215 modifies cost sharing for projects and for periodic renourishment.
- WRDA (2007)—Section 2018 reaffirms policy to participate in renourishment projects. Establishes preference for areas with an existing federal investment, and where impacted by navigation projects or other federal activities.

SOURCE: C. Bronson, USACE, personal communication (2013).

To support project design and enhance federal, state, and local coastal risk reduction efforts, the USACE also conducts coastal risk-related research and develops modeling and sea-level rise mapping tools through its Engineer Research and Development Center and Institute of Water Resources. Through its Floodplain Management Services and Planning Assistance to States programs, the USACE also provides technical and planning assistance to state and local governments to improve flood risk management. In addition, the USACE has some limited ongoing general program authorities to address shoreline erosion, manage sediment resources, and encourage beneficial uses of dredged materials.

As shown in Table 2-2, the average budgets for USACE coastal flooding and storm damage reduction efforts represent a small fraction (ranging from 1.2 to 4.1 percent) of the total Civil Works budget. The vast majority of funds for USACE coastal risk reduction efforts are through emergency supplemental appropriations, passed by Congress in response to specific national disasters (Tables 2-1 and 2-3). Major hurricane risk reduction projects such as the New Orleans Hurricane and Storm Damage Risk Reduction System (HSDRRS), the Hurricane Sandy rebuilding projects, and the North Atlantic Comprehensive Study were funded through emergency supplemental appropriations. Between FY 2008 and FY 2012, $493 million was appropriated for USACE coastal storm risk management efforts through the annual budgeting process, while at least $12.8 billion was allocated for coastal risk projects via supplemental appropriations (Tables 2-2 and 2-3). Further discussions about USACE project planning, authorization, and appropriations process are provided later in this chapter.

TABLE 2-2 Coastal and Inland Flood and Storm Damage Components of USACE Civil Works Budget Appropriations (millions of dollars)

	FY 2008	FY 2009	FY 2010	FY 2011	FY 2012
Coastal flooding and storm damage reduction					
Construction	52	58	59	131	60
Operations and maintenance	16	2	0	58	8
Investigations and other	5	4	11	18	11
Total coastal flooding and storm damage reduction	73	64	70	207	79
Total Inland flooding and storm damage reduction	1,662	1,514	1,796	1,585	1,346
Total USACE Civil Works budget	5,591	5,210	5,449	5,055	5,003

SOURCE: B. Carlson, USACE, personal communication (2013).

TABLE 2-3 Supplemental Funding for USACE Coastal Risk Reduction Projects Since FY 2005

Supplemental	Storm Event Addressed	USACE Funding Appropriated (million $)
P.L. 109-62 (FY 2005)	Hurricane Katrina	400
P.L. 109-148 (FY 2006)	Hurricanes Katrina, Rita, Wilma, Ophelia	2,361
P.L. 109-234 (FY 2006)	Hurricanes Katrina, Rita	3,653
P.L. 110-28 (FY 2007)	Hurricanes Katrina, Rita	1,433
P.L. 110-252 (FY 2008-09)	Hurricane Katrina (+ recent storms)	At least 5,762
P.L. 110-329 (FY 2008)	Hurricane Katrina (+ recent storms)	At least 1,500
P.L. 111-32 (FY 2009)	Hurricane Katrina (+ recent storms)	At least 439
P.L. 113-2 (FY 2013)	Hurricane Sandy	5,081
Total		At least 20,629[a]

[a]An additional $2 billion in supplemental funds were provided between FY 2008 and FY 2010 to address "recent storms," which may or may not include other hurricane events.
SOURCE: B. Carlson, USACE, personal communication (2013).

Federal Emergency Management Agency

FEMA authorities and responsibilities for coastal risk management activities range from direct response to natural disasters to oversight of mitigation programs to administration of the National Flood Insurance Program (see Box 2-2). Additionally, the Homeland Security Act of 2002 tasked the FEMA administrator to lead the nation in natural disaster preparedness, mitigation, response, and recovery. FEMA addresses all types of disasters, but in recent years coastal storm events represent the majority of FEMA's largest disaster relief expenditures. Between 1996 and 2013, out of 15 disasters with FEMA disaster expenditures of at least $500 million, 14 of those were hurricane events, and these major hurricane-related expenditures represented 75 percent of all of FEMA's Disaster Relief Fund expenditures over this period (Figure 2-1; Lindsay, 2014). The following sections briefly summarize the major FEMA programs in the areas of response and recovery, mitigation, and flood insurance. The funding reported for FEMA programs (Table 2-4) is not specific to coastal risk management.

Response and Recovery

After a disaster, FEMA may assist with initial damage assessments and assists in coordinating federal, state, and local response efforts. FEMA also manages several programs to support recovery after a federal disaster declaration.

BOX 2-2
Major FEMA Authorities Related to Coastal Risk Reduction

- National Flood Insurance Act of 1968 (P.L. 90-448)—Created the Federal Insurance Administration and made flood insurance available.
- Flood Disaster Protection Act of 1973 (P.L. 93-234)—Made the purchase of flood insurance mandatory for properties located in special flood hazard areas (the 1 percent chance [or 100-year] floodplain).
- Robert T. Stafford Disaster Relief and Emergency Assistance Act of 1974 (P.L. 93-288, amended in 1988 by P.L. 100-707)—Outlines the means by which the federal government works with local, state, and tribal governments to provide emergency assistance after a disaster.
- Disaster Mitigation Act of 2000 (P.L. 106-390)—Provided the legal basis for FEMA mitigation planning requirements associated with eligibility for pre- or post-disaster mitigation and recovery funds. The Act encourages a more proactive approach to risk mitigation by incentivizing comprehensive and integrated hazard mitigation planning.
- Homeland Security Act of 2002 (P.L. 107-296)—Assigned responsibility to the FEMA administrator to "lead the Nation's efforts to prepare for, protect against, respond to, recover from, and mitigate against the risk of natural disasters, acts of terrorism, and other manmade disasters, including catastrophic incidents; . . . [and] develop and coordinate the implementation of a risk-based, all-hazards strategy for preparedness."

These laws recognize the national interest in disaster prevention and mitigation and provide funding for that purpose, but they also recognize that risk reduction measures need to be coordinated with and, in many cases, undertaken by local communities.

Public Assistance. The Public Assistance Program provides funding to state, tribal, or local governments to repair damaged infrastructure and for debris removal. The program provides at least 75 percent of the costs of eligible projects. Infrastructure is often repaired to pre-flood conditions, unless an effort is made to include mitigation for risk reduction, which would require a benefit-cost analysis.[1]

Individual Assistance. The Individual Assistance Program provides disaster aid directly to individuals, including temporary housing or funding for housing repairs, crisis counseling, and grants to assist with needs not covered by insurance, such as transportation, medical, or funeral expenses.[2]

[1] See http://www.fema.gov/public-assistance-frequently-asked-questions.
[2] See http://www.fema.gov/disaster-process-disaster-aid-programs.

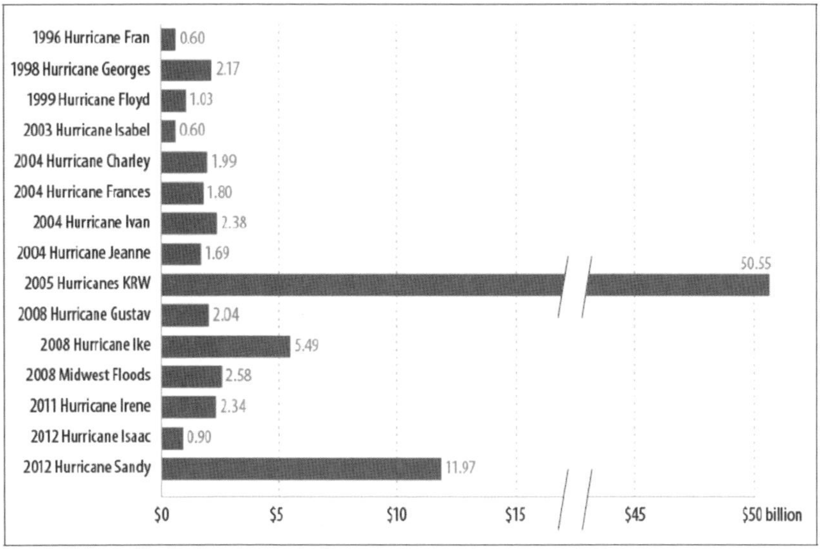

FIGURE 2-1 Disasters between 1996 and 2013 with FEMA expenditures greater than $500 million (actual-year dollars). Amounts reflect FEMA disaster assistance expenditures as of February 2013, and do not included funding provided by other agencies. KRW = Katrina, Rita, and Wilma.
SOURCE: Lindsay (2014).

TABLE 2-4 Components of FEMA Budget Appropriations, including Emergency Supplemental Appropriations (in millions of dollars).

	FY2008	FY2009	FY2010	FY2011	FY2012	FY2013
Disaster Assistance Direct Loan Program						
Annual Appropriation	0.875	0.295	0.295	0.294	0.295	0.27
Emergency Supplemental	98.2	0	0	0	0	300
Disaster Relief Fund[a]						
Annual Appropriation	1,324	1,288	1,600	2,523	7,076	6,653
Emergency Supplemental	10,960	0	5,100	0	6,400	10,914[b]
Mitigation						
Pre-Disaster Mitigation	114	90	100	50	36	24
Flood-related Grants[c]	62	110	79	209	60	120

NOTE: Funding shown here addresses all disasters, not just coastal storms.
[a]Up to 15 percent of Disaster Relief Funds may be spent through the Hazard Mitigation Grant Program.
[b]After sequestration.
[c]Includes the three NFIP mitigation grant programs: the Flood Mitigation Assistance, Repetitive Flood Claims, and Severe Repetitive Loss programs.
SOURCES: DHS (2009a,b, 2010a,b, 2011a,b, 2012b,c, 2013d,e, 2014a,b); Brown (2012), Lindsay (2014).

In FY 2012, the award range was $50-$31,400, with an average award of $2,982.[3]

The Public and Individual Assistance Programs are funded through the Disaster Relief Fund, which receives annual appropriations and emergency supplemental appropriations as needed. For example, the Hurricane Sandy Supplemental (P.L. 113-2) provided $11.5 billion (before sequestration) for the Disaster Relief Fund (Table 2-1). In response to Hurricane Katrina, $43.1 billion was provided for the Disaster Relief Fund in FY 2005 via supplemental funding (Lindsay and Murray, 2011). Funding for the Disaster Relief Fund since 2008 is shown in Table 2-4.

Community Disaster Loans. The Community Disaster Loan program, funded through the Disaster Assistance Direct Loan Program (see Table 2-4), offers low-interest loans to local governments to maintain government functions after a substantial loss of revenue. Under certain conditions of continuing financial hardship, FEMA is permitted to cancel repayment of loans 3 years after the disaster. Brown (2012) reports that FEMA has forgiven $896 million of the $1,326 million loaned to local governments since the program began in 1974.

Mitigation

FEMA has three major programs that provide grants for mitigation efforts: Pre-Disaster Mitigation, the Hazard Mitigation Grant Program, and Flood Mitigation Assistance.

Pre-Disaster Mitigation. The Pre-Disaster Mitigation program is used to fund local, state, and tribal governments and others at the community level to plan for hazards and implement cost-effective risk reduction measures prior to a disaster. The Pre-Disaster Mitigation program provides up to 75 percent of the costs of mitigation activities (or up to 90 percent in small or low-income communities), and the projects are funded through a competitive process (CBO, 2007).[4] Annual appropriations for the Pre-Disaster Mitigation program have ranged from $24 million to $114 million since 2008 (see Table 2-4).

Hazard Mitigation Grants. The Hazard Mitigation Grant Program has provided funds to states after a Presidential disaster declaration to reduce

[3]See https://www.cfda.gov/.

[4]Individual homeowners or businesses are not permitted to apply for funds in this program, although local governments or nonprofits may apply on their behalf. See http://www.fema.gov/pre-disaster-mitigation-grant-program.

or eliminate losses in future disasters. The states then allocate the funds to state and local governments, tribes, and nonprofits for hazard mitigation. Up to 15 percent of total disaster funding for each state may be provided through the Hazard Mitigation Grant Program. At least 25 percent of the costs of eligible mitigation projects must be provided by state or local governments, although funds from the Housing and Urban Development Community Development Block Grants (see discussion below) can be used as the nonfederal cost share.[5] The Hazard Mitigation Grant Program is funded through the Disaster Relief Fund, discussed in the preceding section. Hazard mitigation grants were about 5 percent ($5.2 billion; DHS, 2012a) of disaster relief funds between 2004 and 2012, which totaled over $100 billion (Lindsay, 2014).

Flood-related Grants. The National Flood Insurance Program provides grants through three programs—the Flood Mitigation Assistance, Repetitive Flood Claims, and Severe Repetitive Loss programs—to support mitigation efforts that would reduce the program's future losses. The programs provide funds to state and local governments to support mitigation planning and mitigation projects for structures with repetitive losses that are insured under the National Flood Insurance Program. Recent appropriations have ranged from $60 million to $209 million (Table 2-4).

National Flood Insurance Program

The National Flood Insurance Program (NFIP) was established by the National Flood Insurance Act of 1968 as a means to encourage community-level actions to reduce flood losses and to provide insurance to property owners already at risk of flooding in communities that adhere to floodplain standards. As a condition for participation in the program, local governments agree to adopt and enforce construction standards to reduce potential damage to new or substantially improved buildings in "special flood hazard areas"—the land surface area covered by a 1 percent annual chance (100-year) flood (see Box 1-3).

As of December 2012, the NFIP had over 5.5 million policies in over 21,000 communities in the United States (NRC, 2013). Approximately 19 percent of all policies are discounted—most because the structures were built prior to the development of flood risk maps, and thus received subsidized rates (ASFPM, 2013). The NFIP was not structured to hold sufficient funds for eventual heavy flood losses. Instead, the program was given limited borrowing authority from the federal treasury so insured claims could be paid when losses exceeded premium income (which is set

[5]See http://www.fema.gov/hazard-mitigation-grant-programs-frequently-ask-questions.

to cover the average historical loss year; NRC, 2013). Annual operating losses (premiums minus claims and operating expenses) occurred in many years during the 1970s, 1980s, and 1990s (Michel-Kerjan, 2010), and the NFIP borrowed money periodically but was always able to pay back what it borrowed until 2005, when four major hurricanes hit. The losses from that year (nearly $19 billion) exceeded the total losses of the program since its inception. As of 2013, the debt stood at $30.4 billion (King, 2013). To the degree the program fails to adequately reflect risk in rates and operates at a loss, it has been subsidizing the occupancy of hazardous areas.

In July 2012 the National Flood Insurance Reform Act, known as the Biggert-Waters Act, was passed by Congress and signed into law (P.L. 112-141). The Act phased out subsidized insurance for many of the properties insured under the NFIP, including repetitive loss properties, second homes, businesses, and those that had "grandfathered" rates after a flood risk map update. Under the Act, actuarially based rates would be in effect immediately if a policy lapsed or the residence was sold. These changes mandated by the Biggert-Waters Act would have led to significant rate increases for the approximately 1.1 million NFIP policyholders with subsidized rates.

In response to substantial public outcry over the significant impact of Biggert-Waters on insurance rates and potential impact on property values, Congress adopted the Homeowner Flood Insurance Affordability Act of 2014 (P.L. 113-89), which reinstated discounted rates and repealed the large increase in rates with property sales or lapsed policies. Discounted rates will be increased gradually (up to 18 percent annually for primary residences and up to 25 percent annually for businesses, nonprimary residences, and repetitive loss properties) for all properties until the premiums reach full risk-based rates. The new law also allows for rate adjustments for flood mitigation actions that are not part of the insured structure and implements a surcharge on all policyholders ($25 for primary residences, $250 for all other policies).

In support of the NFIP, FEMA develops and updates flood insurance rate maps (FIRMs), which delineate floodplains with 1 percent and 0.2 percent annual chance of flooding (100- and 500-year floods) and are critical for local and regional planning. FIRMs in many communities are out of date, but FEMA is in the process of updating its rate maps to include high-accuracy, high-resolution topographic data available from current technologies and to reflect recent land-use changes (e.g., NRC, 2009). The current national percentage for new, validated, or updated engineering FIRMs is approximately 64 percent (Paul Rooney, FEMA, personal communication, 2014). FEMA flood rate maps only reflect today's risks and do not forecast the flooding implications of climate change and sea-level rise. However, FEMA was tasked in the Biggert-Waters Act to establish a Technical Mapping Advisory Council to develop recommendations for incorporating climate change science into flood risk assessments.

Department of Housing and Urban Development

HUD coordinates the Community Development Block Grants (CDBG) program, which provides economic development funding for state, local, and tribal governments. Among the program's priorities is funding projects for disaster recovery assistance, especially for low- and moderate-income communities. Disaster Recovery grants can assist with housing buyouts and relocation to safer areas, house repair or replacement, and construction of public infrastructure. As noted previously in this section, these funds can be used as the nonfederal cost-share for the Hazard Mitigation Grant Program. In FY 2013, Congress appropriated $16 billion ($15.2 after sequester) to the CDBG (Table 2-1). The program received appropriations of $16.7 billion in FY 2006 in response to Hurricanes Katrina, Rita, and Wilma and $6.1 billion in FY 2008 for disasters including Hurricanes Ike, Dolly, and Gustav.[6] Although HUD encourages CDGB rebuilding efforts to incorporate preparedness and mitigation, there is no explicit requirement to do so. Of the $1.8 billion in CDBG funds for New York City, approximately $533 million was targeted to enhance disaster resilience (NYC, 2013).

National Oceanic and Atmospheric Administration

NOAA is charged "to advance ocean, coastal, Great Lakes, and atmospheric research and development" (33 USC § 893). Its contributions to coastal risk reduction come from observational data collection and forecasts, inundation modeling and risk reduction decision support tools, coastal zone management, and training. Under the Coastal Zone Management Act of 1972, NOAA administers the Coastal Zone Management Program, which is a partnership between states and NOAA to manage coastal resources, including conservation, recreation, and development. NOAA also administers the Coastal and Estuarine Land Conservation Program, which has a goal of protecting coastal or estuarine areas that have conservation or ecological value.

NOAA research and data collection provide important support for coastal risk management at local, state, and federal levels. Observations of water level, topography and bathymetry, and aero-gravity data are used to create National Weather Service storm surge forecasts and will

---

[6] See http://portal.hud.gov/hudportal/HUD?src=/program_offices/comm_planning/communitydevelopment/programs/drsi.

be used to create inundation maps. The Coastal Services Center supports a Sea Level Rise and Coastal Flooding Impacts Viewer, which is used to visualize future sea-level rise and potential impacts to coastal communities. This tool can also be used for redevelopment planning, especially to visualize the expansion of the flood hazard areas. Additionally, through the Digital Coast program,[7] they provide access to physical, topographic, hazard, and social science data for the nation's coasts.

NOAA's National Ocean Service FY 2013 budget (under a continuing resolution) included $72 million for Coastal Science and Assessment and $114 million for Coastal Zone Management and Services and Coastal Management Grants (NOAA, 2013). The committee does not have information on what percentage of these budgets directly supports coastal risk reduction.

Other Federal Programs

As illustrated in Table 2-1, the federal government provides substantial post-disaster recovery funds to redress the impact of coastal storms on public and private infrastructure through the Small Business Administration, Department of Transportation, Department of Agriculture, Environmental Protection Agency, and other agencies. Additionally, the federal government supports mapping, data collection, tool development, and research to enhance coastal risk management through agencies such as the U.S. Geological Survey, while the Environmental Protection Agency is working to enhance the resiliency of water infrastructure. Some of the key agencies are discussed briefly below.

Small Business Administration (SBA)

The SBA manages the Disaster Loan Program, which can be used for repair or replacement of personal and business property. The low-interest loans are available to homeowners, renters, and businesses and can be used to cover uninsured losses to personal property, including homes, vehicles, and clothing. Additional funding (up to 20 percent of the amount of disaster damage, not to exceed $200,000) can be made available for mitigation measures, such as elevating a home.[8] The Hurricane Sandy Emergency Supplemental Appropriation (P.S. 113-2) included $779 million for the Disaster Loan Program Account.

---

[7]See http://www.csc.noaa.gov/digitalcoast/.

[8]See http://www.sba.gov/content/disaster-loan-program.

Department of Transportation

The Department of Transportation has several major efforts under way with application to coastal risk reduction. The Federal Highway Administration is working to test strategies for assessing transportation infrastructure vulnerabilities to climate change and extreme weather and for improving infrastructure resilience. Additionally, the Federal Transit Administration received Hurricane Sandy Emergency Supplemental Appropriations of $10.8 billion to repair the most impacted transit systems (P.L. 113-2), and as of June 2013, the FTA had allocated $1.3 billion toward mitigation efforts that would enhance the resiliency of transit systems in the region to future disasters (Executive Office of the President, 2013).

U.S. Department of Agriculture (USDA)

USDA's Natural Resources Conservation Service (NRCS) manages several programs to assist landowners after a natural disaster. The Emergency Conservation Program (ECP) provides assistance to restore agricultural land to a productive state after a natural disaster, and the Emergency Forest Restoration Program (EFRP) provides assistance to private, nonindustrial forestland owners to address damage on those lands. The Emergency Watershed Protection (EWP) program supports emergency recovery efforts to prevent erosion and reduce runoff, such as removing debris from clogged stream channels or restoring undermined stream banks.[9] These programs are funded only in response to disasters, and receive no annual appropriations. In response to Hurricane Sandy, Congress appropriated $15 million to the ECP, $23 million to the EFRP, and $180 million to the EWP (Painter and Brown, 2013).

Department of the Interior

U.S. Geological Survey (USGS). Coastal risk- and coastal change-relevant activities are spread broadly across the USGS. USGS mapping programs provide geospatial information including coastal elevation, hydrography, geology, and land cover and land use. Monitoring from the USGS Water Mission Area includes real-time and post-storm observations of storm surge and high water levels and, with national monitoring of riverine flows and water levels, supports forecasts and assessments of coastal inundation hazards. The Ecosystems Mission Area conducts research on the health and vulnerability of coastal wetlands, forests, coral reefs, and species and populations of ecological and commercial concern. The Natural Hazards

---

[9]See http://www.nrcs.usda.gov/wps/portal/nrcs/main/national/programs/landscape/ewpp/.

Mission Area provides tools that can anticipate and respond to hazards, vulnerability, and risk.

The Coastal and Marine Geology Program (CMGP) is conducting a National Assessment of Coastal Change Hazards over multiple timescales, considering hurricanes and extreme storms, long-term shoreline change, sea-level rise, and seacliff erosion. A Web-based portal is being developed to provide access to data, tools, and assessment products for coastal managers and stakeholders. As part of its broad research effort, the CMGP also supports the development of sediment transport models and research on fundamental coastal processes, including regional research studies to provide the process-level understanding necessary to forecast the evolution of coastal systems (e.g., Fire Island, New York; Gulf Barrier Islands, Louisiana and Mississippi). Funding for CMGP base activities, which support coastal change hazards research and products, averaged $13 million per year over the past 5 years (FY 2009-FY 2013) (J. Haines, USGS, personal communication, 2014). In FY 2013, the USGS received emergency supplemental appropriations totaling $41.2 million, including $16 million for the CMGP to enhance data collection, expand vulnerability assessments, conduct regional studies including geological mapping and oceanographic modeling, and to improve the delivery of data, tools, and products (P.L. 113-2).

Fish and Wildlife Service (FWS). The Coastal Barrier Resources Act, enacted in 1982 and reauthorized in the Coastal Barrier Reauthorization Act of 2000, designated relatively undeveloped barriers along the Gulf and Atlantic Coasts, including both private and public lands, as part of the John H. Chafee Coastal Barrier Resources System (CBRS). The Act "encourages the conservation of hurricane prone, biologically rich coastal barriers by restricting Federal expenditures that encourage development,"[10] including federal flood insurance, loans by the Small Business Administration, and subsidies for erosion control, utilities, roads, and bridges. All costs of new development on these lands must be borne by nonfederal parties, although federal funds are still permitted to aid in disaster recovery after a major storm (Salvesen, 2005). The FWS is responsible for administering the Act and estimates that the program saved the federal government over $1 billion between 1982 and 2010. GAO (2007) estimated that 16 percent of the CBRS lands experienced development in spite of the federal funding restrictions, encouraged by strong real estate market pressures, the availability of private insurance, and state and local land-use policies that promote floodplain development. The Department of the Interior is modernizing the maps of the CBRS in the north Atlantic with $5 million funding from the Hurricane Sandy emergency supplemental appropriation (P.L. 113-2).

[10]See http://www.fws.gov/cbra/Act/index.html.

Environmental Protection Agency (EPA)

In addition to helping water and wastewater utilities recover from coastal disasters, EPA has several climate-based initiatives under way related to enhancing the resiliency of water and wastewater infrastructure through the Climate Ready Water Utilities program. EPA recently launched the Climate Resilience Evaluation and Awareness Tool to assist drinking-water and wastewater utility managers in understanding climate change threats, including flooding and extreme weather events, and available adaptive measures.

EPA also supports effective wetlands management through partnerships with state, local, and tribal governments and other stakeholders. EPA established a Coastal Wetlands Initiative to better understand factors related to coastal wetland loss and to protect and restore wetlands. In addition, EPA's Climate Ready Estuaries Program analyzes areas for vulnerability to climate change and develops strategies for adaptation, in association with the National Estuary Program and coastal managers.

Federal Coordination Efforts

There are numerous executive and federal interagency policies that have been adopted to bridge the gaps between federal programs and provide better coordination of federal efforts. Even with these policies, it is recognized that federal agencies and their budgets remain guided primarily by their statutory missions.

The Hurricane Sandy Rebuilding Task Force was a focused effort to improve federal coordination and align federal resources with local recovery and rebuilding priorities. Led by HUD, with membership from more than 20 other federal departments, agencies, and offices, the Task Force made numerous policy recommendations, including a building elevation standard (advisory base flood elevation + 1 foot [0.3 m]) for rebuilding efforts using federal funds (Hurricane Sandy Rebuilding Task Force, 2013).

Many of the federal executive and interagency coordination efforts have been developed under the umbrella of an all-hazards approach, including terrorist acts as well as natural disasters. In response to a Presidential Policy Directive, PPD-8 (2011), a National Preparedness Goal[11] was developed along with five supporting planning frameworks (focused on disaster prevention, protection, mitigation, response, and recovery; DHS, 2013a,b,c) to guide agencies and personnel to operate in a unified and

---

[11]The following National Preparedness Goal was developed in 2011: "a secure and resilient nation with the capabilities required across the whole community to prevent, protect against, mitigate, respond to, and recover from the threats and hazards that pose the greatest risk." See http://www.fema.gov/national-preparedness-goal.

collaborative manner in disaster-related efforts. The National Disaster Recovery Framework, while providing much-needed structure and principles to support coordination, does not attempt to reconcile sometimes differing individual program mandates or authorities. With regard to natural disasters, the framework largely emphasizes mitigation, response, and recovery, rather than addressing the removal of incentives that continue to permit and in some cases encourage development (and redevelopment) that places people and property in harm's way.

Several federal coordination efforts address coastal risk management among other issues. The Federal Interagency Floodplain Management Task Force was authorized by Congress and established under the Water Resources Council in 1975 to develop a proposed framework for a "Unified National Program for Floodplain Management." The Task Force, which consists of 12 federal agencies and is chaired by FEMA, has proposed several such frameworks (FIFMTF, 1986, 1989, 1994). It was reconvened in 2009 after a decade of inaction, and continues to work to unify federal programs on flooding, despite minimal impact of past reports. Recent efforts include guidance on unwise use of floodplains (FIFMTF, 2012) and consensus recommendations for actions by task force agencies and the task force itself to improve floodplain management (FIFMTF, 2013).

President Obama established the Interagency Climate Change Adaptation Task Force, co-chaired by the Council on Environmental Quality (CEQ), the Office of Science and Technology Policy, and NOAA, and including representatives from more than 20 federal agencies (ICCATF, 2011). In October 2009, President Obama signed Executive Order (E.O. 13514) directing the Task Force to recommend ways that federal policies and programs can better prepare the nation for climate change.

More recently, President Obama established an interagency Council on Climate Preparedness and Resilience in November 2013. Among its functions, the Council is tasked to "support regional, State, local, and tribal action to assess climate change related vulnerabilities and cost-effectively increase climate preparedness and resilience of communities, critical economic sectors, natural and built infrastructure, and natural resources" (E.O. 13653). The 2013 executive order also established a task force of state, local, and tribal leaders on climate preparedness and resilience to provide recommendations on "how the federal government can remove barriers, create incentives, and otherwise modernize Federal programs to encourage investments, practices and partnerships that facilitate increased resilience to climate impacts, including those associated with extreme weather."

Other executive orders have been issued to coordinate federal actions and improve the consistency of policies shaping the efforts of federal agencies. For example, Executive Order 11988 (1977) required federal agencies to minimize actions that result in "adverse impacts associated with the occupancy and modification of floodplains and to avoid direct or indirect support of floodplain development wherever there is a practicable alternative." However, executive orders are not always implemented fully and consistently across all federal agencies.

USACE PROJECT PLANNING, AUTHORIZATION, AND FUNDING

The USACE is the primary agency that oversees the planning, design, and construction of projects such as hard structures and beach nourishment to reduce the probability of coastal hazards (e.g., flooding, wave attack). Thus, in addition to the prior discussions of agency responsibilities and budgets, it is important to understand the mechanisms by which USACE coastal risk reduction projects are identified, designed, authorized, and funded. This section summarizes the procedures and criteria that the USACE uses for coastal risk reduction project planning, authorization, and appropriations, in order to provide a context for understanding opportunities for and impediments to improving links with other federal, state, and community risk reduction efforts.

Project Initiation and Planning

Guidance for USACE water resources planning activities, including coastal risk reduction, inland flood risk reduction, navigation, and ecosystem restoration, comes from several sources, but the two most important that are currently in effect are the *Economic and Environmental Principles and Guidelines for Water and Related Land Resources Implementation Studies* (*Principles and Guidelines*; WRC, 1983) and the *Planning Guidance Notebook* (Engineering Regulation 1105-2-100; USACE, 2000a). The *Principles and Guidelines* provide federal agencies (e.g., USACE, the Bureau of Reclamation, the Tennessee Valley Authority) with detailed instructions for evaluating water-related project alternatives. [12] The Planning Guidance Notebook (USACE, 2000a) was developed to provide additional guidance within the USACE in implementing the *Principles and Guidelines*. Additional detailed USACE guidance specific to coastal risk reduction has also been developed (USACE, 2011a,b).

Coastal risk reduction efforts are typically initiated at the local level among a USACE district office, a community or local interest group, and a congressional delegation. If a need is identified, funds are appropriated for the USACE to study the project and determine whether it represents a federal interest. If a federal interest is determined and USACE headquarters

---

[12] *Principles and Guidelines* replaced *Principles and Standards* (WRC, 1973), which served as water-related planning requirements for federal agencies.

approves, a more detailed feasibility study is initiated that includes coordination with state and local entities, public involvement, and development of an environmental impact statement. Feasibility studies for USACE water resources projects typically have each taken about 4.5 years and several million dollars to complete (NRC, 1999). However, USACE headquarters has recently emphasized faster planning through the "3×3×3" initiative, which required feasibility studies to be completed in no more than 3 years, with three levels of vertical team integration, for no more than $3 million (Walsh, 2012). Congress must appropriate half of the funding for the feasibility study, with the remaining funds coming from the local sponsor (NRC, 2004b).

USACE projects, including coastal risk reduction projects, follow six project planning steps identified in the *Principles and Guidelines*:

1. Specify problems and opportunities.
2. Inventory and forecast conditions.
3. Formulate alternative plans.
4. Evaluate effects of alternative plans.
5. Compare alternative plans.
6. Select recommended plan.

This planning process is well described in USACE (2000a) and NRC (2004b).

Benefit-cost analysis serves as the most important decision criterion in project planning (USACE, 2000a). In some cases, specific exceptions from cost-benefit formulation are mandated by Congress, as they were for New Orleans and the Mississippi coast rebuilding efforts.[13] According to the *Principles and Guidelines* (WRC, 1983), the objective of water resources project planning "is to contribute to national economic development [NED] consistent with protecting the Nation's environment. . . ." Thus, projects are designed to maximize NED benefits relative to financial costs, while ensuring that the project does not cause unacceptable adverse environmental impacts. Additional project elements may be necessary to mitigate environmental impacts (USACE, 2011a). Other social effects are also evaluated, but this information rarely influences planning decisions (NRC, 2004b). *Principles and Guidelines* has long been criticized for its narrow focus and failure to factor nonmonetary environmental and social costs or benefits

[13]Projects such as the New Orleans Hurricane Storm Damage and Risk Reduction System and the Mississippi Coastal Improvements Project were designed, based on direction from Congress, without benefit-cost analysis as a basis of decision. Congress directed the New Orleans HSDRRS to be built to provide flood protection against a 100-year storm, and no assessment of benefits versus costs was performed. The MSCIP considered the cost-effectiveness of project components but did not evaluate the design by the benefit-cost ratio.

in project planning and decision priorities, and in the Water Resources Development Act of 2007, Congress directed the administration to revise them. In March 2013, the White House Council on Environmental Quality released the updated *Principles and Requirements for Federal Investment in Water Resources* (CEQ, 2013), which significantly broadens federal interests in water resources projects by stating

> Federal investments in water resources as a whole should strive to maximize public benefits, with appropriate consideration of costs. Public benefits encompass environmental, economic, and social goals, include monetary and non-monetary effects and allow for the consideration of both quantified and unquantified measures.

However, the 1983 *Principles and Guidelines* will not be replaced until 180 days after revisions are completed on the detailed interagency guidelines that are to accompany the *Principles and Requirements*. Congress has also prohibited the USACE from implementing the new *Principles and Requirements* (Explanatory Statement to P.L 113-76). The 1983 *Principles and Guidelines* and the 2013 revised *Principles and Requirements* are discussed further in Chapter 4.

USACE projects are considered "economically feasible" if the national economic development benefits exceed the costs. The major categories of NED benefits that are currently considered in USACE coastal risk reduction projects are (USACE, 2011a)

- reduction in physical damage (including structures, contents, infrastructure, agricultural crops, and land value),
- reduction in nonphysical damages (including income loss, emergency response costs, evacuation, temporary housing, and transportation delays), and
- other benefits, such as increased recreational use, incidental recreation benefits, and land enhancement.

Recreation benefits, however, cannot exceed more than 50 percent of the benefits required for project justification. The value of human lives and well-being is not assessed in current calculations of NED benefits. Although termed "national economic benefits," the beneficiaries may be primarily local in distribution. The planning process performs a separate calculation of "regional economic benefits," but this term encompasses benefits that are transferred from one location to another (i.e., businesses that relocate), and therefore are not net gains from a national perspective.

For coastal storm damage reduction projects, the USACE considers damages from inundation, wave attack, and erosion, and estimates damage prevented by project alternatives. The costs and benefits over a 50-year period of analysis are compared against a scenario of the future without the project alternatives in place. According to current USACE planning guidance (USACE, 2011b), these future scenarios must consider three scenarios of sea-level rise. USACE planning teams also evaluate project alternatives on their economic efficiency, meaning that each increment of a project must produce benefits that exceed costs. Ultimately, the selected plan represents an optimization of the net benefits, both in the overall plan and considering increments of the project, while meeting requirements of completeness, effectiveness, efficiency, acceptability, and compliance with federal, state, and local regulations (USACE, 2011b). Thus, coastal risk reduction projects are *not* designed with a specific level of risk reduction in mind (such as the 1 percent chance [100-year] flood event, see Box 1-3), unless the project is congressionally mandated to do so.

Coastal risk reduction projects that have been constructed by the USACE represent a range of levels of risk reduction, including many beach nourishment and dune construction projects that are built to prevent flooding from storms that have a 3 to 5 percent chance or greater of occurring in any given year (i.e., 20- to 30-year return interval; USACE, 2013c). This outcome is in marked contrast to inland flood risk reduction measures, which often are designed to reduce risks associated with a 1 percent annual-chance (100-year) event or larger for the purpose of alleviating flood insurance requirements for the residents behind the levees. Local sponsors, however, are required to fund the cost difference between the NED-justified design and the 1 percent chance risk reduction level, if the economic analysis does not justify risk reduction to a 1 percent chance event. USACE coastal risk reduction projects can also include measures to reduce the consequences of an event, such as land purchase and relocation, although several past reports have highlighted the USACE's limited emphasis on such strategies (Moore and Moore, 1989; NRC, 1999, 2004b).

Once the preferred project alternative is identified, the USACE prepares a draft feasibility report and environmental impact statement. When finalized by the district office and after public comment and coordination with other federal agencies, the feasibility report is submitted to USACE Headquarters for approval.

Authorization and Appropriations

Following approval of the feasibility study by the Assistant Secretary of the Army (Civil Works) and a subsequent review by the Office of Management and Budget, the USACE feasibility study may be transmitted to Congress for authorization. Water Resources Development Acts (WRDAs) are typically used to authorize water resources projects, although WRDAs have been passed infrequently in recent years—the last three were passed in 2000, 2007, and 2014. Only after authorization can projects be considered for federal appropriations for the federal portion of the projects (see Box 2-3), and funding is not guaranteed. The USACE Civil Works has a backlog of approximately 1,000 projects (including coastal risk reduction projects and other projects, such as navigation, dam and levee safety, and ecosystem restoration) that are authorized but unfunded, representing

BOX 2-3
Cost Sharing for Coastal Risk Reduction

The rules for cost sharing for traditional USACE coastal risk reduction projects vary among the types of USACE projects (Table 2-3-1). Accordingly, the federal government funds 50-65 percent of the construction of most USACE coastal risk reduction projects. However, Congress has provided greater federal funding for the construction of some risk reduction projects after a major disaster (e.g., 100 percent for construction and repair of authorized projects after Hurricane Sandy, 89 percent for the HSDRRS after Hurricane Katrina [USACE, 2012b]). Coastal risk reduction projects that involve beach nourishment are considered continuing construction projects, and thus maintenance costs are shared for the life span of the project. However, for other projects, such as for seawalls or levees, the nonfederal partner is responsible for 100 percent of the operations and maintenance costs. For example, in New Orleans, nonfederal funding is expected to pay for all maintenance costs once the project has been officially completed, even though subsequent "lifts" clearly will be needed to keep the levee heights at their design levels.

TABLE 2-3-1 Cost Sharing Percentages for Various Coastal Risk Reduction Project Scenarios

Project Type	Federal Construction	Nonfederal Construction	Federal O&M
Federal shores	100	0	100
Public or private developed shores with public use	65	35	0[a]
Undeveloped nonfederal public shores	50	50	0[a]
Private developed shores, with private use	0	100	0
Undeveloped private lands	0	100	0

[a]Beach nourishment is considered a continuing construction project, so all renourishment activity costs are shared according to the construction percentages.
SOURCE: Data from USACE (2000a).

about $60 billion (NRC, 2011b; Carter and Stern, 2013). Although positive net benefits are required for project authorization, the benefit-cost ratio used for budgetary prioritization often needs to be above 2.5 to compete for available funds in the President's budget (USACE, 2013g). As discussed previously in this chapter, over 95 percent of USACE funding for coastal risk reduction projects (FY 2008–FY 2012) has also been allocated through separate emergency supplemental appropriations (Tables 2-2 and 2-3).

Supplemental funding provides both advantages and constraints for coastal risk reduction projects. Supplemental post-disaster funds can spur more holistic evaluations and regional perspectives for coastal risk reduction. For example, the New Orleans HSDRRS project was systematically developed for risk reduction and includes a combination of structures that protect the entire region from flooding (USACE, 2013b). Previously, the city's flood risk reduction measures consisted of a series of smaller projects constructed over more than 50 years—"a system in name only" (IPET, 2009). The Mississippi Coastal Improvements Project is another example of a regional study funded by congressional mandate after Hurricane Katrina (USACE, 2009). In the area affected by Hurricane Sandy, the emergency appropriations allow previously authorized but unfunded projects to be constructed and existing projects to be restored to their design level. Key risks remain, however, because the projects being restored were originally designed and analyzed individually, rather than from a comprehensive, systemwide framework. The Sandy supplemental funding is also supporting the North Atlantic Comprehensive Study, which will provide a rare systemwide analysis of opportunities for coastal risk reduction. These regional studies and projects, which were funded at full federal expense, represent the exception rather than the norm for USACE coastal risk efforts. A major constraint of supplemental funding is that it tends to be reactive in nature, providing funding for risk reduction and resilience efforts only after an area has been impacted by a coastal storm. The funding provides little support for other areas of the nation at risk from future storms.

Challenges and Constraints Within the USACE Planning and Authorization Process

The USACE's coastal risk reduction planning and design process, described in the preceding sections, has evolved over the last 50 years or so to meet changing needs and priorities within the United States. During this time, USACE activities were also expanded to include environmental restoration, and the cost of authorized projects to meet diverse USACE missions has far exceeded the appropriated funding. This section examines the adequacy of the existing USACE planning and authorization process to address the nation's increasing coastal risks from severe storms and rising sea levels.

Local Interests and Regional Planning

Although local engagement is essential in terms of ensuring that all work being considered is coordinated with local stakeholders, such interests often originate from relatively narrow segments of the at-risk area rather than the region as a whole. Within the current USACE planning process, it is far easier to consider individual projects within a specific geographic area that have a single purpose, such as beach nourishment, than to develop a comprehensive plan for coastal risk reduction. Regional comprehensive planning requires engagement of multiple local sponsors, each contributing to funding and planning a project beyond its jurisdiction. Such efforts require intense engagement and agreement among multiple local governments—some of which might not fare as well under a systemwide risk reduction effort as they might in a narrowly focused project. A comprehensive, regional coastal risk reduction project considering the full range of vulnerabilities (i.e., not just beach nourishment but also including back-bay flooding and urban areas) would take much longer to plan and could result in a project with a substantially higher cost. Comprehensive coastal risk reduction studies would also require specific authority and funding. Overall, these issues make it difficult to address coastal risk reduction on a regional scale within the USACE process.

Funding and Prioritization

Only a small fraction of annual USACE appropriations are directed toward coastal storm damage reduction projects (see Table 2-1), and competition for these funds is fierce. Instead, coastal risk is primarily addressed by the USACE via emergency supplemental funding *after* a disaster has occurred. As previously discussed, the result is that the nation is reactive to disasters, rather than proactive in addressing priorities at a national scale.

The nation also lacks a mechanism to weigh coastal risk reduction investments for large coastal cities from a national perspective. As noted in Chapter 1, eight large cities in the United States rank among the world's top 20 in terms of estimated potential average annual losses from coastal flooding (Hallegatte et al., 2013) and numerous others face significant risks. Addressing coastal risks in densely developed urban areas requires extensive investments at systemwide scales, likely costing billions of dollars per city. In 2013, Mayor Bloomberg announced a plan to reduce New York City's coastal risk costing at least $20 billion (NYC, 2013). Existing USACE annual appropriations are insufficient to address these challenges, and the project-by-project authorization process does not allow the Congress to take stock of the nation's coastal risks and plan a strategy to reduce them.

The capacity to support operations and maintenance of coastal risk reduction projects remains an additional challenge. Aside from beach nourishment projects, which are considered continuing construction projects for their life span, minimal funding is typically allocated for operations and maintenance of hard structural risk reduction projects (e.g., levees, surge barriers; see Table 2-1), because such funding is borne by local sponsors under current guidance. If localities are unable or unwilling to maintain existing coastal risk management projects to their original designs, local communities may be exposed to elevated risks and much lower benefits from the original federal investment.

STATE, LOCAL, AND NONGOVERNMENTAL RESPONSIBILITIES IN COASTAL RISK REDUCTION

As described above, most of the federal coastal risk reduction programs are designed to be implemented by or in collaboration with state and local governments that have primary responsibility and authority over planning, economic development, and land use. In addition, numerous reports have emphasized the importance of stronger private–public sector partnerships and community collaboration to strengthen community resilience and reduce risk (e.g., NRC 2001, 2010). States develop hazard mitigation plans to be eligible for FEMA pre- and post-disaster funding. Under the Disaster Mitigation Act (Box 2-2), states and localities support development of local mitigation plans and provide technical assistance to local governments. States administer FEMA's Hazard Mitigation Grant Program and establish funding priorities consistent with their hazard mitigation plans. Local communities submit individual project applications to the state, coordinate with participating homeowners and businesses, and manage implementation of the approved projects.

State and local governments can implement building codes with minimum design and construction requirements that reduce the vulnerability of new structures in high-hazard areas. Model codes are usually developed and periodically revised by nongovernmental organizations (e.g., the International Code Council). Many but not all of the states affected by Hurricane Sandy have adopted the international building code and the international residential code. Based on a recommendation in the Hurricane Sandy Rebuilding Strategy (Hurricane Sandy Rebuilding Task Force, 2013), the Mitigation Framework Leadership Group (MitFLG) is currently working to encourage state, local, and private-sector adoption of the most recent (2012) version of these international model codes. Inconsistent adoption or enforcement of codes at the state and local levels can leave communities vulnerable.

The Coastal Zone Management Act (CZMA) of 1972 (discussed above) provides a framework for federal-state cooperation on coastal hazard management, land use, and development. One of the main objectives of the CZMA is to "minimize the loss of life and property caused by improper development in flood-prone, storm surge, geological hazard, and erosion-prone areas and in areas likely to be affected by or vulnerable to sea level rise" (16 USC § 1452). With federal funding support and technical assistance under the CZMA, state and local governments develop coastal hazard management plans and conduct projects that address coastal hazards, such as revising construction setback regulations and mapping shorelines to identify high-risk erosion areas.

State and local governments can develop and implement their own plans for coastal risk reduction. The State of Louisiana's Coastal Protection and Restoration Authority prepared its 2012 Coastal Master Plan to address the massive erosion of shorelines and wetlands that the state has experienced. The plan included hundreds of possible remedies for land loss and coastal risk reduction, including nonstructural measures, and called for extensive investment in coastal work in Louisiana for the next 50 years, with a total cost of $50 billion dollars (CPRA LA, 2012). State and local governments also partner with the USACE to develop and fund coastal risk reduction projects, per cost-sharing requirements (Box 2-3).

At the state and local levels, actions to consider risk reduction beyond that required by the programs described above have largely taken the form of adaptation plans. Examples include multisector climate adaptation plans in California (CNRA, 2009), a regional climate change compact in southeast Florida (SFRCCC, 2012), and ongoing development of state climate change adaptation plans in Maryland and Delaware. These plans have been informed by the recommendations of the federal Interagency Climate Change Adaptation Task Force (ICCATF, 2011). There are also numerous efforts in North Carolina and other states to map future shoreline change to evaluate how sea-level rise will impact flooding and risk management strategies (Burkett and Davidson, 2013). However, at the local level, it is challenging to maintain long-term climate change adaptation and resilience planning programs, although state and federal involvement can help sustain local efforts.

Nongovernmental organizations such as the Association of State Floodplain Managers and Coastal States Organization also provide policy recommendations that identify how federal efforts can better support state and local risk management and best practices for states. Other nongovernmental organizations such as The Nature Conservancy (along with federal, university, and private-sector partners) provide decision support tools for vulnerable coastal areas.[14]

ALLOCATION OF RISK, RESPONSIBILITY, REWARDS, AND RESOURCES

A major impediment to U.S. coastal hazard management is the misalignment of risks, rewards, responsibilities, and resources associated with coastal development and post-disaster recovery. If the risks, responsibilities, rewards, and resources are primarily borne by a limited number of agencies, as they are in the Netherlands, coastal risk management becomes more straightforward. However, in the United States, risks, rewards, responsibilities, and resources are each borne by different entities motivated by different objectives.

The *rewards* of coastal development flow to developers, engineers, architects, and builders, as well as local and state governments in the form of contracts, profits, and tax revenue. Rather than concluding simplistically that communities are acting recklessly in allowing development along the coast, it is important to recognize that local officials are acting rationally to the extent that development makes local economic sense, provides tax revenues, results in greater local employment and, in some cases, reflects the preservation of historical and cultural community values. It is ordinarily in the best interest of the property owner, developer, builder, and municipality to undertake construction regardless of future public risk and other externalities. Although local governments also bear some of the risk associated with storm damage to coastal development, other beneficiaries, such as developers and builders, evade such risks.

Risks associated with coastal development are borne by individual home and business owners (particularly those without flood insurance) and federal, state, and local governments (and their taxpayers) that fund disaster relief and recovery programs. Risks are also borne by coastal residents who face economic and social disruption after a severe coastal storm. However, behavioral and cognitive factors hinder individuals and organizations from taking appropriate risk reduction actions (Kunreuther, 2006). One limiting factor is the human tendency to be more accepting of risks associated with natural hazards (Slovic, 1987). Many people view natural hazard risks, especially those posed by low-probability/high-consequence events, as facts of life and acts of nature that are often inexplicable and cannot be completely avoided. The importance of risk reduction efforts is likely to be eclipsed, for both public officials and the general public, by

[14]See http://www.coastalresilience.org/tools.

more immediate and pressing concerns. The federal government also bears the risks of insuring flood losses through the National Flood Insurance Program, which offers discounted rates to some policyholders, contributing to the program's $30 billion shortfall (King, 2013).

Responsibilities associated with sound land-use planning decisions fall primarily to local governments. Although the framework for U.S. emergency management provides that local communities are encouraged to mitigate, prepare for, and respond to disasters, in reality the incentives have been relatively weak. Behavioral research has shown that people are more likely to favor investments that generate immediate benefits than those that yield long-term benefits that accrue probabilistically (Kunreuther, 2006). Thus, it is generally much easier to elicit a sense of concern from public officials and the public for issues such as unemployment, economic development, crime, and traffic congestion, which affect the public almost daily. Meanwhile, localities depend on local tax revenues, enhanced by expanded development, to fund schools and other essential public services. The Stafford Act currently requires states and communities to develop and update hazard mitigation plans, but these plans are rarely incorporated into local economic development or land-use master plans.

Following major disasters, many look to the federal government for *resources* to fund emergency response, individual and community post-disaster assistance, redevelopment programs, mitigation efforts, and coastal storm damage reduction projects. In recent years, the federal government has borne a larger share of the costs associated with major hurricanes (Table 1-5). These efforts shift risk to federal taxpayers, thereby encouraging more intensive development and rebuilding in high-risk areas.

These government services may actually promote a phenomenon referred to as moral hazard (Mileti, 1999; Kunreuther, 2008). A moral hazard describes the possibility that individuals and organizations will take more risks when they believe that they will be protected from the consequences of their decisions. In the case of hazards, there is concern that individuals and local governments will continue to pursue floodplain development and avoid spending scarce resources on disaster preparedness and mitigation based on a belief that the federal government will bail them out (Platt et al., 2002). The Hurricane Sandy Emergency Supplemental Appropriations included little guidance or requirements that the expenditures would result in making communities, people, and property more resilient to future storms. Much of the $48 billion in funding (after sequester) was provided to support response or recovery programs at full federal expense, removing local and state funding requirements that can serve as disincentive to simply rebuilding regardless of long-term consequences. The U.S. Commission on Ocean Policy (2004) noted the need to coordinate the efforts of all coastal risk management agencies to reduce inappropriate incentives and to "establish clear disincentives to building or rebuilding in coastal high-hazard zones."

CONCLUSIONS

The lack of alignment of risk, reward, resources, and responsibility as it relates to coastal risk management leads to inefficiencies and inappropriate incentives that serve to increase the nation's exposure to risk. Developers, builders, and state and local governments reap the rewards of coastal development but do not bear equivalent risk, because the federal government has borne an increasing share of the costs of coastal disasters. The resulting moral hazard leads to continued development and redevelopment in high-hazard areas.

Responsibilities for coastal risk reduction are spread over a number of federal, state, and local agencies, with no central leadership or unified vision. Multiple federal agencies play some role in coastal risk management, and each agency is driven by different objectives and authorities. No federal coordinating body exists with the singular focus of mitigating coastal risk, although several efforts are under way to increase coordination.

The vast majority of the funding for coastal risk-related issues is provided only after a disaster occurs, through emergency supplemental appropriations. Pre-disaster funding for mitigation, preparedness, and planning is limited, and virtually no attention has been given to prioritization of coastal risk reduction expenditures at a regional or national scale to better prepare for future disasters. Thus, efforts to date have been largely reactive and mostly focused on local risks, rather than proactive with a regional or national perspective. Also, although the federal government encourages improved community resilience, only a small fraction of post-disaster funds are specifically targeted toward mitigation efforts.

Few comprehensive regional evaluations of coastal risk have been performed, and the USACE has no existing institutional authority to address coastal risk at a regional or national scale. Given the enormous cost of coastal disasters within the United States, which are rising, improved systemwide coastal risk management is a critical need within the federal government. Under the current planning framework, the USACE responds to requests at a local level on a project-by-project basis, and several major urban areas remain at significant risk. Barriers effectively prohibit the USACE from undertaking a comprehensive national analysis of coastal risks and strategies to address them, unless specifically requested and funded by Congress.

3

Performance of Coastal Risk Reduction Strategies

Until the latter part of the 1900s, the use of hard structures in coastal areas (sometimes termed coastal armoring), was the preferred method for reducing the effects of waves, storm surge, and erosion. However, over recent decades, U.S. Army of Corps of Engineers (USACE) beach nourishment and dune building outnumbered hard structures in terms of both the number and miles of projects constructed (see Figure 1-9). Additional approaches use natural or restored habitats to help reduce the impact of waves and storm surge, and/or building design and nonstructural land-use strategies to reduce the consequences of a hazardous event. This chapter discusses the various coastal risk reduction strategies and reviews what is known about their proven performance, including their economic costs and benefits and environmental effects.

Isolating the economic value of coastal risk reduction projects in comparative studies of protected and unprotected areas is difficult to accomplish because of lack of available data. In this chapter, the committee reports available relevant data, but it did not examine the categories of losses reported or attempt to ensure consistency among the types of damages included in loss calculations—a step that would be necessary in a rigorous analysis of costs and benefits, but one that is beyond the scope of this study. Most of the quantitative data reported in this chapter on damages, however, comes from the USACE, which follows specific protocols for reporting data on damages prevented (Comiskey, 2005). Each locality is unique with respect to exposure to storms, value and age of structures, and amount of development on the oceanfront versus the bay side, complicating direct comparisons of protected and unprotected communities (USACE, 2000a).

Although coastal risk reduction provides social benefits by protecting cultural and historic resources and allowing residents and business owners to feel more secure about their personal safety and assets, these benefits are difficult to quantify, and studies documenting the proven social costs and benefits of specific coastal risk reduction strategies are scarce. Therefore, the chapter does not address social impacts of various risk reduction strategies, although it in no way is meant to lessen the importance of these issues.

IMPLEMENTATION OF COASTAL RISK REDUCTION APPROACHES

Of the roughly 3,700 miles (6,000 km) of coastline along the U.S. East and Gulf Coasts (or over 45,000 miles [72,000 km] of tidal shoreline[1]), the USACE has constructed coastal risk reduction projects on over 640 miles (1,000 km) (see Appendix B for a listing of USACE projects). For the USACE projects, information is generally available about the types of projects and date of construction, but the committee was unable to obtain information on the level of protection provided by each of the projects.[2] Additional coastal risk reduction projects have been constructed by state or local governments or private parties, but this information is held by state and local governments and has not been centrally compiled. London et al. (2009) reported that many of the coastal states have hard structures (e.g., seawalls, bulkheads, revetments; see Figure 3-1) that were built in the 1950s during an intense period of coastal development (see Figures 1-6 and 1-7). In recent decades, however, such structures have been more difficult to implement and are sometimes prohibited at the state level. Most of the Corps efforts related to coastal risk mitigation within the last two decades have focused on beachfront areas, with a heavy reliance on beach nourishment as the primary means of coastal risk reduction (Figure 1-9; Appendix B). This includes many beaches in New Jersey, Florida, and other East and Gulf Coast states. Approximately 40 percent of Florida's coastline and 17

[1]The term coastline represents "a general outline of the seacoast" while the tidal shoreline represents a more detailed measure, including offshore islands, sounds, bays, and the tidal portions of rivers and creeks. See http://nationalatlas.gov/articles/mapping/a_general.html.

[2]The USACE used to report design "levels of protection" for projects based on recurrence intervals (i.e., 40 year, 200 year) when deterministic evaluation procedures were used (see Box 1-2), but the data have not been compiled for constructed coastal projects. Also, design levels of protection reflect conditions upon construction, and dune profiles and sea level change over time. Thus, the USACE now describes projects in terms of the specific design, damages reduced, and residual risks (B. Carlson, USACE, personal communication, 2014).

FIGURE 3-1 An eroding beach adjacent to State Road A1A in Flagler County, Florida, recently under study by the USACE for additional hurricane and storm damage risk reduction.
SOURCE: USACE (2014).

percent of New Jersey's coastline are protected by a USACE storm damage reduction project (C. Bronson, USACE, personal communication, 2013).

The results of a telephone survey of coastal managers conducted by a group of researchers in South Carolina offer evidence that many states use strategies to reduce the consequences of a coastal storm, such as hazard zoning, building elevation, land purchase, or setbacks (see Table 3-1; London et al., 2009). Most of these measures attempt to discourage development in undeveloped, hazard-prone areas or encourage people to abandon their use. The strategies reportedly used by the most states were building elevation requirements, fixed setbacks, and land purchase, with only a few states reporting utility-line limits, abandonment, low-density development, relocation, or rolling setbacks. Not including building elevation, nonstructural strategies were generally viewed as difficult or somewhat

TABLE 3-1 Percentage of Coastal States Using Consequence Reduction Strategies, by Region

Modification of Development Tools	All East and Gulf States (18 states)	Northeast (5 states)	Mid-Atlantic (5 states)	Southeast (4 states)	Gulf of Mexico (4 states)
Building elevation	89	100	80	100	75
Fixed setback	83	100	100	75	50
Land purchase	78	100	100	50	50
Hazard zoning	72	100	80	25	75
Hazard reconstruction limits	61	60	60	100	25
Rolling setback	39	60	40	25	25
Relocation	39	40	20	50	50
Low-density development	33	60	20	50	0
Abandonment	33	60	20	25	25
Utility/service-line limits	28	40	0	25	50

NOTE: States were divided as follows: Northeast (Maine, New Hampshire, Massachusetts, Rhode Island, Connecticut), Mid-Atlantic (New York, New Jersey, Delaware, Maryland, Virginia), Southeast (North Carolina, South Carolina, Georgia, Florida [both East and Gulf coasts]), Gulf of Mexico (Texas, Louisiana, Mississippi, Alabama).
SOURCE: Adapted, with permission, from London et al. (2009).

difficult to implement (Table 3-2; London et al., 2009). Nonstructural strategies were given significantly lower scores by the responding states for ease of implementation compared with either beach nourishment and vegetation strategies ("soft stabilization") or hard structural strategies.

Natural and nature-based approaches to coastal risk reduction include using or expanding dunes, salt marsh, mangroves, reefs, and seagrass to mitigate flooding and erosion associated with wave action or storm surge. The role of intact or restored ecosystems in providing for flood defense and enhancing resiliency to natural disasters along the East and Gulf Coasts has received substantial attention over the last decade and is an area of active research (e.g., Arkema et al., 2013; Duarte et al., 2013; Renaud et al., 2013; Temmerman et al., 2013). In contrast to heavily engineered structures, nature-based initiatives may have ecosystem benefits that extend beyond mitigation of coastal risk, including improved water quality, provision of fisheries habitat, and nutrient sequestration. Moreover, unlike engineered structures, biogenic habitats may have some potential for vertical accretion (e.g., oysters; Rodriguez et al., 2014), which could mitigate effects of sea-level rise (Lenihan, 1999). Living structures can also adapt to sea-level rise by shifting inland, in contrast to permanently emplaced hard structures. Of the 142 USACE projects listed in Appendix B, only 4 include natural

TABLE 3-2 Reported Ease of Implementation of Coastal Risk Reduction Strategies by Region (1= difficult, 4 = easy)

	Northeast (5 states)	Mid-Atlantic (5 states)	Southeast (4 states)	Gulf of Mexico (4 states)
Hazard Reduction Measures: Hard structures				
Seawall	1 (difficult)	1.2	1.62	0.75
Bulkhead	1.25	1.8	1.88	1.75
Jetty	1.25	1.6	2.88	1
Revetment	1.25	2.2	2.62	1.5
Groin	1	1.6	3	1.25
Hazard reduction measures: Nature-based strategies				
Beach nourishment	3.25	3.2	2.75	1
Bulldozing/scraping	2	2.8	1.25	2
Dune addition	3	2	3.5	1
Vegetation	3.25	3.8	4 (easy)	1.67
Consequence reduction measures: Building design and nonstructural land-use strategies				
Building elevation	3.12	2.3	1.75	1
Fixed setback	1.75	1.4	1.5	0.75
Land purchase	0.75	1.4	1	1
Hazard zoning	1.25	0.6	0.75	0.75
Hazard reconstruction limits	0.75	1.4	1.25	1
Rolling setback	0.25	0.6	0.75	0.5
Relocation	1	0.2	1	0.5
Low-density development	1.25	0.4	0.5	0.25
Abandonment	0.75	0.2	0.75	0.25
Utility/service-line limits	0.25	0.6	0	0.25

SOURCE: Adapted, with permission, from London et al. (2009). Note that any value below 1 indicates that at least one respondent answered "not applicable," which was scored with a 0. Respondents were the same states listed in Table 3-1.

systems as an explicit focus of the coastal risk reduction project (termed "ecosystem restoration" in Appendix B).

Louisiana's Comprehensive Master Plan for a Sustainable Coast (CPRA LA, 2012) and the Mississippi Coastal Improvements Program (USACE, 2009) both endorse the "multiple lines of defense" concept, which utilizes a combination of nature-based approaches and hard structures to maximize storm surge risk reduction. As depicted in Figure 3-2, barrier islands shelter the coast from storm waves, and wetlands offer storm surge and wave attenuation, reducing the storm surge and wave heights against landward hardened structures. The effectiveness of each "line" in this defense system and how they work together to provide risk reduction is a subject of active research.

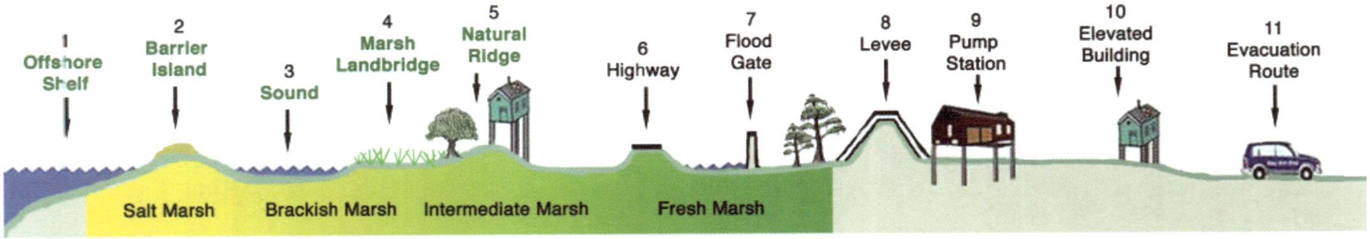

FIGURE 3-2 Various components of the multiple lines of defense approach for Louisiana.
SOURCE: Reprinted, with permission, from Pontchartrain Basin Foundation. Available online at www.mlods.org.

Historically, there has been little emphasis placed on risk reduction measures for large urban areas along the coasts. However, the widespread flooding and loss of life in New Orleans due to Hurricane Katrina in 2005 and the extensive damage throughout New York City from Hurricane Sandy in 2012 demonstrate the impacts for metropolitan areas that are not adequately protected against storm surge and wave attack. Aerts et al. (2014) estimate that without risk reduction measures in place, New York City could have annual flood losses averaging $174 million per year.

PERFORMANCE OF CONVENTIONAL HARD STRUCTURES FOR COASTAL HAZARD REDUCTION

The most commonly deployed hard structures to address coastal flooding are shore-parallel walls (e.g., seawalls, revetments, bulkheads) that are designed for deployment where there is high wave energy, and levees (dikes) and floodwalls are usually employed against flooding in the absence of significant wave action. Storm surge barriers are also used across river mouths or inlets to keep surge from propagating up rivers and estuaries (see Chapter 1). Other structures deployed to prevent coastal erosion, such as offshore breakwaters, do not provide significant risk reduction from storm surges, but are used to reduce loss of sediment from the beach and to reduce storm-induced erosion of protective dunes and levees. Of the 144 USACE coastal storm damage reduction projects listed in Appendix A, approximately half include hard structures of some form.

Seawalls and Shore-Parallel Structures

Seawalls and other shore-parallel structures (such as revetments and bulkheads; Figure 3-3) are built to reduce coastal risks to infrastructure where the natural beaches and dunes have been eliminated or significantly restricted and where other risk reduction options are prevented by lack of space or sediment. These hard structures are also intended to hold the shoreline position by reducing wave attack, storm surge, and associated erosion of landward areas. Although all shore-parallel structures absorb wave energy, different structure types are designed for different coastal settings (see Figure 3-3). Given this report's focus on reducing risk associated with hurricane storm surge, this section focuses on seawalls.

Overall Effectiveness for Reducing Flood Damage

Seawalls have a long history of being used to prevent flood damage and casualties during severe coastal storms, and the Galveston Seawall is one the earliest examples (Wiegel and Saville, 1996). Galveston, Texas, was

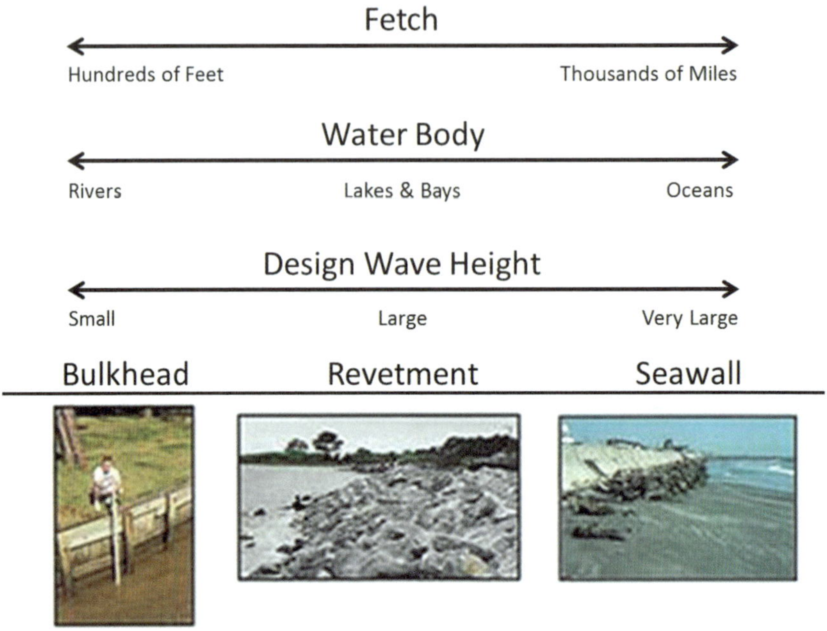

FIGURE 3-3 Types of shore-parallel structural risk reduction strategies.
SOURCE: Douglass and Krolak (2008).

devastated by a Category 4 hurricane that made landfall on September 4, 1900, killing more than 6,000 inhabitants and causing more than $111 billion (2013 dollars) in damage (Blake et al., 2011). It remains the deadliest natural disaster in U.S. history. The original seawall was completed in 1904, with several subsequent extensions; the present length is 10.6 mi (17 km), more than three times the original length. The seawall has protected the city from 13 hurricanes (Kraus and Lin, 2009). The 1915 hurricane, for example, was considered as severe as the 1900 event and had a particularly long duration. The city still saw similar economic damage (Pielke et al., 2008, and the seawall itself sustained major damage. A total of 11 people died; the reduction in fatalities compared with the 1900 hurricane is often credited to the existence of the seawall (Davis, 1961). Minor wave overtopping occurred during Hurricane Ike in 2008, and Kraus and Lin (2009) report that bayside flooding, as opposed to ocean-side flooding over the seawall, led to inundation in the city. Areas without dunes or seawalls were heavily damaged or destroyed.

During field observations to quantify damage after Hurricane Sandy, Irish et al. (2013) reported that a buried relic seawall lessened the hurricane's impact relative to adjacent areas (Figure 3-4). They found that pre-storm beach and dune profiles were similar between two neighboring boroughs (Bay Head and Mantoloking), but that a 0.75-mile-long (1.2-km) stone seawall constructed in 1882 lay buried below the Bay Head dune. Field observations showed that water levels in the two communities were similar, but the number of homes that were damaged or destroyed was significantly higher in the community without the seawall. Through numerical modeling, they estimated that the existence of the seawall, although not capable of preventing overtopping, decreased wave forces by a factor of two. Other factors, such as differences in the width of the barrier island and proximity to existing or former inlets, may have also contributed to the contrasts between boroughs.

Economic Costs and Documented Returns to Date

Data are sparse on the documented economic benefits of seawalls or other shore-parallel structures. In 1904, the originally specified length of the

FIGURE 3-4 Relic seawall in Bay Head, New Jersey, that was buried beneath a dune and uncovered during Hurricane Sandy.
SOURCE: Photo courtesy of Jennifer Irish. Available online at http://www.nsf.gov/news/news_images.jsp?cntn_id=128545&org=NSF.

Galveston Seawall was completed at a cost of $1.6 million (approximately $40 million in 2013 dollars) (USACE, 1981). The seawall protected the city not only against the unnamed hurricane in 1915, but also against Hurricane Carla in 1961 (Davis, 1961) and Hurricane Alicia in 1983. If the seawall had not been in place, another $234 million (2013 dollars) in damage could have occurred from the 1983 storm alone (K. Bohnam, USACE quoted in Associated Press, 1984).

Linham et al. (2010) reported that the cost of a vertical seawall ranges from $0.6 million/mile ($0.4 million/km) to $44 million/mile ($27 million/km), with price affected by factors including the design height, anticipated wave loadings, and construction materials. The UK Environment Agency (2007) reported average costs of $4.3 million/mile ($2.7 million/km) (in 2013 dollars).

Impacts on Adjacent Areas

The impact of seawalls on adjacent areas is a many-decades-old debate; reviews include Kraus (1988), Pilkey and Wright (1988), and Kraus and McDougal (1996). Possible adverse effects include offshore steepening of the beach profile, intensified local scour at the toe or end of the seawall, transport of sand to a substantial distance offshore, downdrift erosion, and delayed post-storm recovery (Dean, 1986). Other possible adverse effects include reduction of the beach width fronting the seawall and acceleration of the beach erosion rate (Basco, 2006). Kraus and McDougal (1996) note that because seawalls are often built in areas with chronic beach erosion, passive versus active erosive forces need to be distinguished. Passive erosion refers to "tendencies which existed before the wall was in place" (Griggs et al., 1991, 1994) and active erosion refers to the "interaction of the wall with local coastal processes" (Kraus and McDougal, 1996). Dean (1986) provides an assessment of commonly expressed concerns relating to coastal armoring (Table 3-3).

Coastal armoring is generally used in areas that are already experiencing erosion. According to Dean (1986), it is likely that coastal armoring will cause an increase in erosion on adjacent unprotected beaches because hard structures prevent the protected beach from providing sediment to the rest of the coastal system. Eroding beaches will continue to erode, while the structures prevent the progressive landward movement of the protected shoreline. However, Dean (1986) concludes that there are no observations or physical arguments to support the concern that coastal armoring causes accelerated erosion seaward of a protected beach, causes the beach profile to steepen dramatically, or delays post-storm recovery.

TABLE 3-3 Assessment of Some Commonly Expressed Concerns Related to Coastal Armoring

Concern	Assessment	
Coastal armoring placed in an area of existing erosional stress causes *increased* erosional stress on the beaches adjacent to the armoring.	True	By preventing the upland from eroding, the beaches adjacent to the armoring share a greater portion of the same total erosional stress.
Coastal armoring placed in an area of existing erosional stress will cause the beaches fronting the armoring to diminish.	True	Coastal armoring is designed to protect the upland, but does not prevent erosion of the beach profile seaward of the armoring. Thus an eroding beach will continue to erode. If the armoring had not been placed, the width of the beach would have remained approximately the same, but with increasing time, would have been located progressively landward.
Coastal armoring causes an acceleration of beach erosion seaward of the armoring.	Probably False	No known data or physical arguments support this concern.
An isolated coastal armoring can accelerate downdrift erosion.	True	If an isolated structure is armored on an eroding beach, the structure will eventually protrude into the active beach zone and will act to some degree as a groin, interrupting longshore sediment transport and thereby causing downdrift erosion.
Coastal armoring results in a greatly delayed post-storm recovery.	Probably False	No known data or physical arguments support this concern.
Coastal armoring causes the beach profile to steepen dramatically.	Probably False	No known data or physical arguments support this concern.
Coastal armoring placed well back from a stable beach is detrimental to the beach and serves no useful purpose.	False	In order to have any substantial effects on the beaches, the armoring must be acted upon by the waves and beaches. Moreover, armoring set well-back from the normally active shore zone can provide "insurance" for upland structures against storms.

SOURCE: Dean (1986).

Environmental Impacts

Hydrodynamic and morphologic response of a beach to a wall and the ecological effects depend on the position of the wall on the beach relative to breaking waves and swash, with fewer impacts the farther landward the wall is from the shoreline (Weggel, 1988; Plant and Griggs, 1992; Hearon et al., 1996; Dugan and Hubbard, 2006; Dugan et al., 2008). Habitat losses in front of the walls increase as the beach narrows over time because sediment moves out of the area and cannot be replaced by erosion of the protected land (Dugan and Hubbard, 2006; Dugan et al., 2008). The high intertidal zone that is most strongly altered by a newly placed wall can be a key location for spawning habitat for several species of fish and horseshoe crabs; the same areas may be critical for foraging predators (Dugan et al., 2008; Jackson et al., 2008).

Species abundance and diversity, for both invertebrates and shorebirds, can be significantly lower along fronting walls (Dugan et al., 2008; Sobocinski et al., 2010). As the system evolves, the entire intertidal zone of the beach can be eventually eliminated, leaving only subtidal habitat and an intertidal zone composed of construction materials such as wood, boulders, or concrete (Nordstrom, in press). The wall itself can still provide habitat, but can decrease the area of intertidal habitat, reduce the capability of many fauna in settling on it, limit the distribution of specific species, and place species that would live at different positions in the intertidal gradient closer to each other (Bulleri and Chapman, 2010; Chapman and Underwood, 2011). These habitats favor spread of introduced and invasive species.

New structures with relatively homogeneous surfaces lack the crevices, small fractures, pits, and holes that can occur on natural rocks or on structures that have been in place for a long time and have undergone weathering (Moschella et al., 2005). Benthic communities on human structures become more similar to natural rock with increasing age, although communities in these two types of habitat can remain distinct after several decades (Moschella et al., 2005; Burt et al., 2011). In a limited number of studies, structures with surface complexity (rough or pitted rather than smooth and flat) have been associated with increased species diversity (Moschella et al., 2005; Chapman and Underwood, 2011).

Structures can be built or modified to retain, restore, or add natural, semi-natural, or artificial landforms and habitats (Moschella et al., 2005; Browne and Chapman, 2011; Chapman and Underwood, 2011; Nordstrom, in press). Larger structures can be altered to enhance habitat by providing low-angle slopes and greater surface complexity (Chapman and Underwood, 2011)—for example, building rip-rap revetments rather than vertical walls. Great potential exists for altering designs to reduce hard structures' environmental impacts and enhance habitat value, but more research is required to achieve the most desirable assemblage for a given region (Chapman and Underwood, 2011).

Increased shoreline hardening within estuaries is a serious environmental concern. In Mobile Bay, coastal armoring has increased by approximately 0.5 percent per year since 1955 (Douglass and Pickel, 1999) with approximately 38 percent of the shorelines protected by bulkhead or riprap (Jones et al., 2009). A similar pattern has been documented for Chesapeake Bay (Bilkovic and Roggero, 2008). The loss of coastal vegetation decreases the nursery value of nearshore areas (Bilkovic and Roggero, 2008), and also affects nutrient regulation and denitrification. Enhanced scour at the base of seawalls can deepen the sediment base and result in lower photosynthesis, further reducing the ecological value of shallow estuarine waters.

Levees and Floodwalls

Unlike seawalls, coastal levees in the United States are generally not designed to resist the direct attack of high-energy waves. Instead, levees are often located landward of large areas of salt marsh that dissipate wave energy or along rivers or waterways inland from the coast to reducing flooding associated with coastal storm surge. In the Netherlands, large coastal levees with concrete revetment blocks or asphalt cover have been built directly along the coast to prevent flooding and wave attack associated with storm surge. To prevent landward flooding, levees either extend all the way around the land to be protected or are tied into higher upland that prevents inundation from other sides.

Floodwalls (Figure 1-7) are used in areas where there is insufficient land for the large footprint of levees. They need to be stable enough to withstand an overtopping storm surge, including the scour that can occur on the upland side of the floodwall after overtopping.

Overall Effectiveness for Reducing Flood Damage

Coastal levees can be very effective at protecting homes and other infrastructure if appropriately designed for the anticipated water levels. After Hurricane Katrina, Congress authorized and funded the Hurricane and Storm Damage Risk Reduction System (HSDRRS) to reduce risks of storm surge from a 1 percent annual-chance (100-year) event for greater New Orleans and parts of southeast Louisiana. The $14.4 billion project includes 350 miles (560 km) of levees and floodwalls, 73 pumping stations, 3 canal closure structures with pumps, and 4 gated outlets. The HSDRRS was tested for the first time in August 2012 for surge protection from Hurricane Isaac, and the USACE reported that the entire HSDRRS operated as designed (USACE, 2013c).

There have, however, been notable coastal levee failures. During Hurricane Katrina, a combination of waves and surge overtopped the Mississippi Gulf River Outlet levee, leading to major levee breaching and flooding of the Lower Ninth Ward and St. Bernard Parish. Floodwalls in several parts of New Orleans also failed for a variety of geotechnical reasons at levels below the project design, flooding large swaths of the city (IPET, 2007).

Economic Costs and Documented Returns to Date

Planning-level cost information for a coastal levee designed for a 1 percent chance (100-year) event is reported by USACE (2013f) as $8.4 million/mile ($5.2 million/km) for the total estimated first construction cost. The total estimated first construction costs for a floodwall designed for a 1 percent chance event was $28 million/mile ($17 million/km). However, because coastal levees and floodwalls can be designed to withstand varying degrees of storm surge and wave attack, costs will vary widely with design specifications. Although operations and maintenance costs nationwide have not been compiled, RAND Corp. (2012) reported that several New Orleans levee districts will incur approximately $3 million to $5 million/year in operations and maintenance costs for their portions of the HSDRRS, and for the Orleans Levee District will approach $20 million/year. The committee was unable to find publicly available information on economic returns from coastal levee projects based on actual storm events.

Impacts on Adjacent Areas

Large-scale levee construction changes the hydrodynamics of an estuarine system and can decrease the accommodation space for floodwaters, increasing the likelihood of flooding in adjacent areas that lack the levee's protection. Increased water levels in unprotected communities could have significant and immediate economic and life-safety impacts. There are varying levels of impact. While the New Orleans HDSRRS provided protection for areas within the system during Hurricane Isaac, some low-lying communities outside the system experienced devastating storm surge and rainfall flooding (USACE, 2013c), which led to public concern that the HDSRRS increased the amount of flooding to areas beyond it. In an assessment of the HDSRRS's performance during Hurricane Isaac, the USACE found that water-level changes outside the system that could have been due to construction of the HDSRRS were 0.4 ft (0.12 m) or less (USACE, 2013c). However, a proposed project that was meant to keep water from entering Lake Pontchartrain, Louisiana, via levees and gates was shown (through models) to increase storm surge levels along the Mississippi coast by up to 3 ft (0.9 m) (Ben C. Gerwick, Inc., 2012). Adverse impacts on adjacent unprotected areas can be minimized by using smaller ring levees to protect only the most critical human infrastructure.

Environmental Impacts

Levees fundamentally change flow conditions above and below the structure. Even when water is allowed to flow through gates in levees, salinity values and the flux of nutrients can be modified to a point that entirely different biological communities develop on either side of the structure and fish migration patterns are altered. Construction of new levees without provision for natural water exchange results in the elimination of wetlands. Levees also prevent the delivery of waterborne sediment to formerly inundated areas, contributing to subsidence and reducing the potential to keep pace with sea level rise.

Levees in the coastal environment have little direct environmental value, but there is increasing interest in modifying the design of coastal risk reduction features to increase ecological functions and reduce the overall impacts (Day et al., 2000). The "rich levee" concept (Dijkman, 2007) proposes to increase biodiversity within wide coastal levees (as used in the Netherlands) by providing diverse habitats within the structural design, including vegetated shore-parallel ridges and carefully selected armoring material that could provide habitat for diverse flora and fauna.

Storm Surge Barriers

Overall Effectiveness for Reducing Flood Damage

There are only a few storm surge barriers in the United States, although major systems installed abroad demonstrate their efficacy. The Eastern Scheldt barrier in the Netherlands (completed in 1986) and the Thames barrier in the United Kingdom (completed in 1982) have prevented major flooding. Lavery and Donovan (2005) note that the Thames barrier, part of a flood risk reduction system of barriers, floodgates, floodwalls, and embankments, has reliably protected the City of London from North Sea storm surge since its completion.

Four storm surge barriers were constructed by the USACE in New England in the 1960s (Fox Point, Stamford, New Bedford, and Pawcatuck) and a fifth in 1986 in New London, Connecticut. The barriers were designed after a series of severe hurricanes struck New England in 1938, 1944, and 1954 (see Appendix B), which highlighted the vulnerability of the area. The 1938 hurricane damaged or destroyed 200,000 buildings and caused 600 fatalities (Morang, 2007; Pielke et al., 2008).

The 2,880-ft (878-m) Fox Point Barrier (Figure 1-8) stretches across the Providence River, protecting downtown Providence, Rhode Island. The barrier successfully prevented a 2-ft (0.6-m) surge elevation (in excess of tide elevation) from Hurricane Gloria in 1985 and a 4-ft (1.2-m) surge from Hurricane Bob in 1991 (Morang, 2007) and was also used during Hurricane Sandy. The New Bedford, Massachusetts, Hurricane Barrier consists of a 4,500-ft-long (1372-m) earthen levee with a stone cap to an elevation of 20 ft (6 m), with a 150-ft-wide (46-m) gate for navigation. The barrier was reportedly effective during Hurricane Bob (1991), an unnamed coastal storm in 1997 (Morang, 2007), and Hurricane Sandy. During Hurricane Sandy, the peak total height of water (tide plus storm surge) was 6.8 feet (2.1 m), similar to the levels reached in 1991 and 1997. The Stamford, Connecticut, Hurricane Barrier has experienced six storms producing a surge of 9.0 ft (2.7 m) or higher between its completion (1969) and Hurricane Sandy. During Hurricane Sandy, the barrier experienced a storm surge of 11.1 ft (3.4 m), exceeding that of the 1938 hurricane (USACE, 2012).

Economic Costs and Documented Returns to Date

The cost and maintenance requirements of storm surge barriers are likely to limit them to intensively developed areas, where they are or would be key elements in risk reduction strategies (Jiabi et al., 2013; Walsh and Miskewitz, 2013). The committee was able to obtain data on the construction costs and estimated economic returns of three USACE storm surge barriers. The Stamford Hurricane Barrier was completed in 1969 at a cost of $14.5 million (approximately $100 million in 2013 dollars).[3] The USACE estimates that the project has prevented $96 million (2013 dollars) in flood and coastal storm damages, as of September 2013. Construction of the New Bedford Hurricane Barrier was completed in 1966 at a cost of $18.6 million (approximately $140 million in 2013 dollars).[4] The USACE estimates that $52 million (2013 dollars) in flood damage was prevented due to the operation of the hurricane barrier through September 2013. The Fox Point Hurricane Barrier was completed in 1966 at a cost of $15 million (approximately $110 million in 2013 dollars),[5] and the USACE estimates that $5.8 million (2013 dollars) in damages has been prevented by the project through 2013 (N. Frankel, USACE, personal communication,

[3]See http://www.nae.usace.army.mil/Missions/CivilWorks/FloodRiskManagement/Connecticut/Stamford HurricaneBarrier.aspx. Note that inflation adjustments for the projects listed here were based on the midpoint of the construction period.

[4]See http://www.nae.usace.army.mil/Missions/CivilWorks/FloodRiskManagement/Massachusetts/New Bedford.aspx.

[5]See http://www.nae.usace.army.mil/Missions/CivilWorks/FloodRiskManagement/RhodeIsland/FoxPoint.aspx.

2014). Note that maintenance costs are not reflected in the above costs and could represent major additional costs over the life span of these projects.

Impacts on Adjacent Areas

As with large levee projects (discussed in the section above), storm surge barriers can affect the hydrodynamics of the surrounding areas. For example, the slowdown in current velocities via floodgates or storm surge barriers will reduce the sediment transport potential and channel-scouring ability of the currents, leading to increased deposition rates in some parts of the basins and decreased deposition in the tidal delta (Saeijs and Geurts van Kessel, 2005). Eelkema et al. (2013) observed an increase in wave-driven features and a steady erosive trend in the Eastern Scheldt delta in the Netherlands after construction of the storm surge barrier.

Environmental Impacts

Gates and barriers restrict the evolution of coastal-dependent species, because they reduce the effect of ocean or estuarine processes along tidal channels. Restrictions to flooding by saline waters limit the spatial distribution of coastal species, and species that depend on extreme flood events to prevent the proliferation of competitors and predators are likely to be adversely affected by gate closures. Reduction of tidal amplitude due to the Eastern Scheldt barrier has decreased the amount of saltwater marshes and caused shoaling of tidal creeks within the estuary (Saeijs and Geurts van Kessel, 2005). The number of floodgate closures due to sea-level rise is likely to increase dramatically in the future (Carbognin et al., 2010), potentially increasing adverse effects on coastal habitats.

PERFORMANCE OF NATURAL AND NATURE-BASED APPROACHES

Natural landforms and habitats provide a first line of defense in reducing the risk of wave damage, overwash, and flooding in the coastal zone. They can also have positive effects as part of coastal risk reduction projects, assuming that there is enough space for a sufficiently functional natural system between the coastal hazard and the area to be protected. Truly natural landforms are most likely to effectively provide risk reduction when used in concert with nonstructural land-use approaches or for portions of the shoreline dedicated to conservation or restoration, because the inherent dynamism of natural processes (see Box 3-1) does not provide a stable system. If landform stability is required, natural landforms will need to be augmented to reduce their mobility and susceptibility to overwash through

BOX 3-1
Role of Natural Dynamism in the Context of Coastal Risk

Beaches are wave-built and thus do not provide barriers at an elevation that restricts overwash or surge during severe storms. Instead, the natural flood prevention function is provided by sand dunes, an integral component of natural sandy shore systems that are linked to the beach in cycles of sediment exchange. Sediment moves from the dune to the beach through erosion by storm waves, followed by the gradual delivery of sand from the beach to the dune by wind action and swash processes. This sediment exchange is a sign that the beach and dune are functioning naturally, and the dune is at the proper location, given the sediment budget and wave and wind climate. Under natural conditions, dunes are inherently dynamic, and the risk reduction they provide against overwash and flooding varies spatially and temporally, with periods of overwash alternating with periods of dune building that provide temporary stability. There is diversity among the dune systems of the East and Gulf Coasts. In some areas along the Gulf of Mexico, dunes are neither high nor extensive, which is thought to be due to reduced winds and high sand moisture.

The overwash process delivers sediment to the bayshore of barrier islands, allowing them to migrate landward through time while maintaining sediment volume. Overwash that does not reach the back bay builds up the height of the barrier islands, allowing them to keep pace with sea-level rise. Inlets are periodically created across low portions of the barrier islands, migrate alongshore, and may close naturally due to delivery of sediment from updrift. While open, inlets deliver sediment inland, creating flood tidal deltas and providing sediment for bay beaches and salt marsh to form back-barrier habitat. Natural dynamism thus is not a threat to maintenance of barrier islands and spits under natural conditions; however, it is a threat to human facilities with a fixed position on inherently mobile landforms.

The rate of change in natural environments could increase in the future from high rates of sea-level rise or from lessening sediment supply. This could lead to less or even no protective benefits during storms—for example, through barrier island breakup or complete submergence ("drowning in place") (McBride et al., 1995; FitzGerald et al., 2008).

time. These "nature-based" approaches primarily include beach nourishment; dune building; and conservation, creation, or restoration of salt marsh, seagrass, mangroves, and oyster and coral reefs. The sections below focus mostly on these human-modified nature-based systems and their role in coastal risk reduction rather than on truly natural approaches, although these are mentioned where appropriate. As discussed in the sections below, a nature-based system does not necessarily have the same environmental properties as its natural counterparts. They may also have impacts—both positive and negative—on other uses of the coast, such as navigation, ecosystems and fisheries, and recreation and tourism.

Understanding the geological setting and/or human use of a project area is critical to evaluating hazard-related problems and suitable solutions. Although coastal flooding is an issue on both the high-relief coasts of Maine and Massachusetts and the barrier island coasts along much of the East and Gulf Coasts, the solutions may be different. Geological structure affects topography, shoreline orientation, sediment erodibility, and the location, volume, and grain-size characteristics of sediment delivered to the coast, which, in turn, create distinctive natural environments. Past human uses are important because they reflect the historical context and present levels of development, and these legacy effects determine both the suitability of mitigating measures and the size and cost of implementation. Additionally, matching coastal risk reduction strategies to the geological setting preserves the habitat while minimizing disruptions to traditional recreational and economic use of the land. For example, placing gravel on a sandy barrier would disrupt traditional use of a coast with significant beach tourism.

Beach Nourishment and Dune Building

Beaches, when combined with sand dunes, reduce the risks of storm surge–related wave attack and flooding on barrier islands and the mainland. Natural beaches can be widened and dunes enlarged through beach nourishment projects to reduce coastal storm risks for developed areas, although such actions come with both benefits and costs. Beach nourishment and dune building are currently significant parts of the USACE's strategy for coastal risk reduction (see Figure 1-9); thus, the committee focused substantial attention on discussion of these options and their environmental benefits and adverse impacts.

Overall Effectiveness for Reducing Flood Damage

Beach nourishment projects provide fill sediment to counteract and/or repair erosion while increasing the distance of coastal infrastructure from the surf and swash zones and providing space and a source of sand that favor formation and protection of dunes. The coastal risk reduction value of dunes is in their height and volume. The height of the crest determines its value as a barrier against wave attack and flooding, while a sufficient sediment volume allows dunes to survive wave erosion during storms, maintaining the integrity of the crest. Naturally evolving dunes are often too low, narrow, mobile, or discontinuous to protect immediately landward infrastructure, and so they are often augmented by stabilizing them and increasing their height and volume through emplacement of sand-trapping fences or vegetation or by depositing sediment in fill operations.

It is often difficult to separate the amount of risk reduction provided against storm surge and wave attack caused by the extra width of a nourished beach from the effects of the dune superimposed on it. Post-storm damage surveys in New Jersey, North Carolina, and South Carolina reveal that together, increased beach width and dunes reduce flooding and storm damage (Houston, 1996a). The nor'easter storm of March 1962 in New Jersey caused extensive overtopping and elimination of dunes along many communities, but not in areas where property owners had implemented programs to build up dunes (USACE, 1962, 1963). Pre- and post-storm surveys following Hurricane Sandy in New Jersey in 2012 revealed that both beach width and dune height were critical in preventing breaches and overwash, even in locations that were not nourished (Coastal Research Center, 2013). A well-maintained dune in Seaside Park survived the storm, while dunes in nearby municipalities that did not have aggressive dune-building programs suffered overwash, leading to the loss of many homes.

Hurricane Hugo in South Carolina in 1989 revealed the value of high, wide dunes. Post-storm surveys at Myrtle Beach revealed less storm surge penetration and damage to structures behind the widened beach and enlarged dunes than adjacent areas. Homes landward of large dunes received little damage directly from waves and water levels, while low dunes were removed down to the planar beach level. Dunes up to 23-ft (7-m) high and 50ft (15-m) wide were cut back by waves but maintained their integrity (Stauble et al., 1991). All dunes less than 50-ft (15 m) wide were completely eroded (Thieler and Young, 1991), and nine breaches occurred in barrier islands and spits in places where dunes were small or nonexistent (Stauble et al., 1991).

The value of beach fill and dune building in protecting against moderate-energy hurricanes was dramatically revealed in North Carolina as a result of Hurricanes Bertha and Fran in 1996 and Dennis and Floyd in 1999. Hurricanes Dennis and Floyd caused no damage to buildings behind three USACE-constructed dune projects, but damaged or destroyed over 900 buildings located outside the project dunes (Rogers, 2007). Federal Insurance Administration claims for damage caused by storm surge and wave attack and overwash resulting from Hurricanes Bertha and Fran revealed far less damage to structures in locations protected by USACE beach nourishment projects than in adjacent unprotected locations (USACE, 2000a).

The impact of a storm on beaches, dunes, and the upland depends on both storm severity and landform characteristics. Regional distinctions in beach and dune characteristics result in different susceptibility to overwash and flooding, even under the same storm wave and surge characteristics (Sallenger, 2000). Building dunes to increase levels of risk reduction over natural levels may require different standards depending on the region, although the general concepts remain the same. Small dunes are readily eroded and overtopped and are not effective in preventing landward flooding, regardless of the region in which they occur. Beach nourishment and dune construction are not generally well-suited for application to most major urban centers or areas with large port and harbor facilities because of the space requirements and the level of risk reduction desired.

Beaches and dunes on the ocean shore protect against water levels and waves coming from that side, but they do not prevent back-bay flooding from water that passes through coastal inlets and bay mouths (USACE, 2000b). The problem of bayside flooding and the inability of existing coastal risk reduction projects to prevent it were dramatically demonstrated during Hurricane Sandy in New Jersey and New York in October 2012 (USACE, 2013d). Dunes constructed on barrier islands, however, could reduce the possibility for overwash or breaching, potentially lessening the likelihood of bay flooding.

Economic Costs and Documented Returns

The costs of beach nourishment and dune building depend upon the size and location; methods of borrow, transport, and placement of materials; estimated renourishment requirements; and ultimately, how the coast evolves. The level of risk reduction afforded by a beach nourishment project also varies over time, as the beach and dune are eroded by natural processes, requiring periodic renourishment (varying by location). USACE (2013f) reports the costs of beach restoration that reduces the risk of flooding to 1 percent annual chance as $18 million/mile ($11 million/km), with renourishment costs of $6.3 million/mile ($3.9 million/km). However, many beach nourishment projects are not designed to this level and might therefore be constructed at lower costs.

A general case for the economic value of beach nourishment and dune-building projects can be made by noting the great economic costs of losses due to coastal storms and the documented reduced damages behind enhanced dunes discussed in the preceding section (USACE, 1962, 1963, 2000b, 2013d; Stauble et al., 1991; Houston, 1996a; Coastal Research Center, 2013; Rogers, 2007). However, reliable economic data to quantify the benefits of these projects are lacking. Reported damages associated with past hurricanes (see Appendix A) provide a sense of the potential economic benefits if coastal storm damage reduction projects could significantly reduce damages.

The economic benefits of beach nourishment extend well beyond coastal risk reduction. Houston (2013) estimates that about half of Florida tourists are beach tourists, who in 2012 spent about $36 billion directly (including $12.6 billion from international tourists), including more than $6 billion in local, state, and federal taxes. A wider beach supports more business to the beach communities and subsequent community growth

(Jones and Mangun, 2001; Houston, 2013). Finkl (1996) argues that adding the economic returns of beach nourishment projects for tourism to returns for coastal risk reduction makes a compelling case for nourishment as an economic investment.

Viewpoints of the value of beach nourishment depend on whether the observer has primary interest in damage reduction, perceived retention or enhancement of property values, recreation, or environmental benefits (Camfield, 1993; Nordstrom, 2005). Debate occurs on some of the differences in viewpoints (e.g., Houston, 1991; Pilkey, 1992), especially cost-effectiveness.

Impacts on Adjacent Areas

Nourished sand is subject to the same erosional pressures as the original beach, and can move onshore, offshore, or alongshore. Sand transported alongshore moves to downdrift beaches or to an inlet or offshore shoal. Sedimentation in tidal channels at inlets as a consequence of beach renourishment can be considered a detrimental side effect of longshore transport. Most nourishment operations are designed to minimize longshore transport of fill (Beachler and Mann, 1996; Houston, 1996a; Bocamazo et al., 2011). Unwanted sedimentation is often addressed using terminal structures, which can be linked to sand bypass operations, which transfer sediments to eroding segments downdrift of stabilized inlets, or backpass operations, which recycle sediments back to eroding portions of the project area. Sand transported onshore as part of the overwash process is often put back onto the beach post-storm. This procedure hinders the natural migration of the beach, but preserves the sand at the nourished beach.

An increase in the sediment budget downdrift of fill areas enhances the likelihood for landforms to evolve, increasing topographic diversity in a way that is more natural than by direct nourishment. Movement of fill sediment from developed areas to adjacent natural areas helps create wider beaches and larger dunes in those areas and reduces the likelihood that undeveloped enclaves adjacent to coastal risk reduction projects will be weak links in regional plans.

Adverse Environmental Impacts

Many studies have been conducted on the environmental impacts of beach nourishment, but knowledge gaps still exist (Peterson and Bishop, 2005). Beach nourishment can have both positive and negative effects on environmental resources, but negative effects dominate in the short term. Sand placed on the beach during nourishment typically results in immediate mass mortality of sand-dwelling organisms (Peterson et al., 2006), and thus loss of ecosystem function. Newly deposited sediments on beaches typically support similar populations of aquatic sediment-dwelling (infaunal) animals within 1 to 2 years after renourishment (Gorzelany and Nelson, 1978; Leewis et al., 2012), although a thorough investigation of recovery is hampered by issues related to the adequacy of post-nourishment monitoring and sampling designs (Peterson and Bishop, 2005). The long-term, cumulative ecological implications of repeated burial from large-scale nourishment projects are still unknown (Lindeman and Snyder, 1999; Posey and Alphin, 2002; Speybroeck et al., 2006). A critical strategy to reduce ecological impacts is to restrict the timing of dredge and fill operations to winter months, when impacts on biota are reduced. Conducting nourishment in winter may still allow spring-summer recruitment and colonization by infaunal invertebrates under certain conditions (Peterson et al., 2006).

The effects of substrate disturbance on subtidal organisms that use stable hard-bottom or low-energy conditions are largely unknown (Nelson, 1989, 1993), but hard-bottom fauna and flora (including corals and seagrasses) adjacent to sandy beaches may be adversely impacted by additional sedimentation or poor water quality from renourishment. These impacts can be significant, but detecting physical changes as a result of nourishment is difficult, given the background natural variability (Jordan et al., 2010).

Long-term ecological recovery is dependent on a number of factors, including sediment quality and quantity, the nourishment technique, and the size and place of the nourishment (Speybroeck et al., 2006). Use of sediment that differs from native materials (e.g., grain size and shape, compaction, shear resistance, moisture retention) can change the habitat characteristics and evolutionary trend of the beach (Jackson et al., 2002). The longest recovery times seem to be where the grain-size characteristics of the fill materials are poorly matched to the natural beach (Reilly and Bellis, 1983; Rakocinski et al., 1996; USACE, 2001). Alterations of sediment characteristics can influence nesting and hatching success of turtles (Crain et al., 1995; NRC, 1995). Sediment on the active foreshore can be quickly reworked to resemble native sediment, but sediment within the inactive fill area on the backshore will retain the characteristics it had when emplaced, which can delay faunal recovery (Schlacher et al., 2012). Placing poorly sorted sediment on the beach will lead to removal of sand by wind and leave a coarse shell or gravel lag surface that resists aeolian transport and restricts natural dune evolution (van der Wal, 1998; Marcomini and López, 2006; Jackson et al., 2010). Silts and clays increase turbidity during placement or when reworked from the fill during storms. Mismatched fine fill sediments can inhibit burrowing of species from all intertidal zones (Viola et al., in press), while sediment that is coarser or shellier than native sediment may have detrimental effects on recovery of the natural benthic invertebrate community and foraging habitat for other species, including surf fishes and shore birds (Peterson et al., 2006, 2014; Van Tomme et al., 2012; Manning et al., 2013).

The scale of the project and the volume of sediment used also impact the extent of ecological effects. Although difficult to express quantitatively, in general, the more sediment placed between the active shoreline and human infrastructure, the greater the potential for new habitats to form and survive storm wave attack (Nordstrom et al., 2012). However, nourishing a beach at an elevation higher than a natural beach (Figure 3-5) often creates a vertical scarp that restricts movement of fauna and

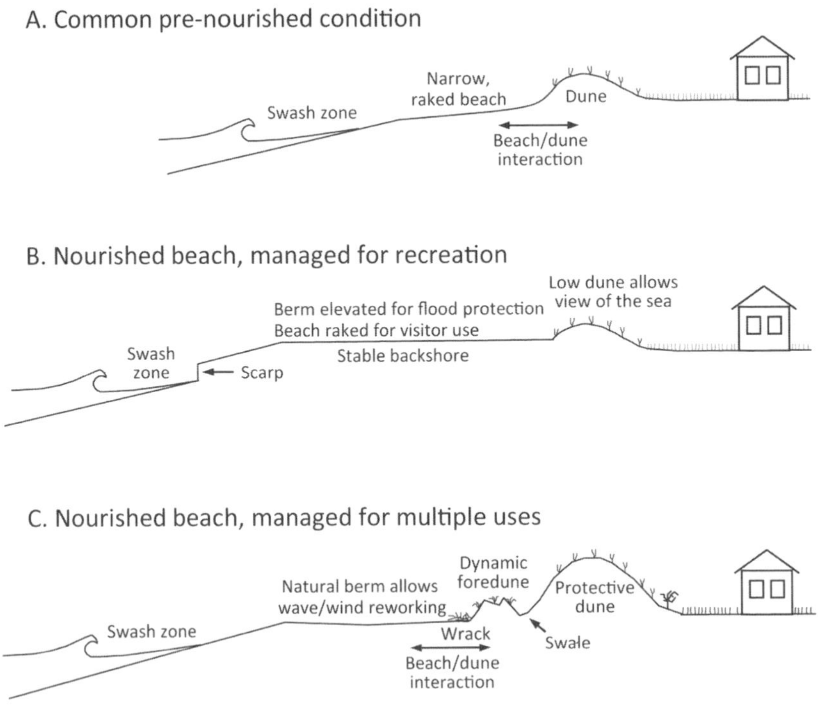

FIGURE 3-5 Beach nourishment scenarios for coastal risk reduction and implications for beach function. Nourishing a beach at the height of a natural berm (C) or even lower allows natural processes to form the final berm (Dean, 2002) and is more compatible with natural habitat requirements. The seaward portion of the fill area could be a dynamic natural zone dedicated to nonconsumptive human uses, while the landward portion could be composed of a relatively static protective dune. The volume of fill sediment would be the same in (B) and (C), but the implications for wildlife would differ greatly.

impedes natural reworking of the backshore by waves and wind (Jackson et al., 2010; Convertino et al., 2011). Large beach nourishment projects can have greater short-term adverse ecological impacts than projects that are small or introduce sediment at a slower rate (Bilodeau and Bourgeois, 2004; Schlacher et al., 2012). Some species depend on gradual recolonization of fill from the edges, and so shorter nourished beaches should recover more quickly, although strategies are available to mitigate these impacts, including slowing the rates of beach fill and leaving small areas unfilled to function as feeder sites for biota (FWS, 2002; Bishop et al., 2006; Schlacher et al., 2012). Many of these project modifications could reduce the impacts of beach nourishment on beach invertebrates, shorebirds, and surf fishes at greater costs per unit of sediment emplaced (Manning et al., 2013).

An important adverse environmental effect of building unnaturally high dunes on barrier islands is that by protecting against overwash, the dunes prevent natural accretion processes that help the island sustain itself. Barrier islands and spits are prevented from keeping pace with sea-level rise or from reestablishing now-rare dynamic habitats, such as washover fans that are favored environments for piping plovers (Maslo et al., 2011; Schupp et al., 2013; see Box 3-1).

The process of offshore sediment extraction to support beach nourishment projects also has important ecological impacts. Dredging of sand borrow areas can decrease organism abundance, total biomass, and the number and composition of species, as well as the average size of the dominant species (USACE, 2001; Brooks et al., 2006). Abundance of some benthic species may recover within a year; recovery of biomass and species diversity may occur within a year or somewhat longer, but changes in biomass composition may take longer. Long-lasting and deep borrow pits may attract a different biological community than occurred before and restricted water movement may create hypoxic or anoxic conditions (USACE, 2001). To reduce impacts, dredging can be restricted to winter months and small, unmined refuge patches can be left within borrow areas (Cutter et al., 2000; Minerals Management Service, 2001; Hobbs, 2002). However, only a small effort has been directed at assessing impacts on biological assemblages from sand removal, especially with respect to ecosystem function (Brooks et al., 2006; Michel et al., 2007). Likewise, long-term impacts have not been addressed.

Offshore borrow pits also have potentially important physical effects on wave regime. Using wave models, Kelley et al. (2001) show that borrow sites can impact the sediment transport regime at the shoreline if the borrow areas are nearshore, large, and deep. Repeated use of the same sediment source area, resulting in deeper offshore pits, would lead to a greater eventual impact on the shoreline. Because most offshore sites are far removed from active sediment sources, such as the nearshore zone and river mouths, these sites are not likely to be refilled with sand by natural processes, resulting in a permanent change to the offshore bottom. In addition, turbidity effects may occur kilometers away from the borrow area (Newell et al., 2004; Pezzuto et al., 2006).

Current and Potential Environmental Benefits

Beach nourishment also has potential environmental benefits, although some would necessitate changes to existing practices and design to achieve. Many of these changes would be relatively easy to accomplish. The biggest potential environmental benefit associated with nourishment is its ability to maintain and/or reestablish habitats and human–nature relationships lost through past coastal development. Habitat loss may be the single biggest threat to biodiversity and ecosystem integrity (Pimm and Raven, 2000; Malanson, 2002; Seabloom et al., 2002) and is common on eroding sandy coasts where human development restricts the onshore migration of beaches and dunes (De Lillis et al., 2004; Feagin, 2005; Schlacher et al., 2007; Dugan et al., 2008). Reestablishing coastal habitats can be accomplished as a direct goal of nourishment or as an unintended outcome, as occurred with the reappearance of the seabeach amaranth (*Amaranthus pumilus*), piping plovers (*Charadrius melodus*; Nordstrom et al., 2000), and tiger beetles (*Cicindela dorsalis*; Fenster et al., 2006). Widening beaches creates staging areas for migratory birds, provides space for nests far from the high-tide line (where they are less subject to flooding and disturbance by people), and decreases competition for nesting resources (Doody, 2001). A nourished beach may increase the cross-shore gradient of physical processes and provide additional habitat (Freestone and Nordstrom, 2001). These benefits are demonstrated by the beach nourishment project conducted at Ocean City, New Jersey, where much of the habitat was restored by natural processes after beach raking and sand fence deployment ceased (Nordstrom et al., 2011). These often unintended consequences of beach nourishment reveal definite potential for incorporation of environmental benefits as specific design outcomes of future projects (Nordstrom, 2008).

Beach nourishment can be used to enhance habitat for specific macrofauna (such as organisms that forage in the swash zone or build nests on beaches) or to allow for accumulation of wrack (seaweed, seagrasses) that may ultimately serve as a food resource for a number of species. Size and sorting characteristics of sediment can be selected to increase spawning rates and egg development for species that make use of the beach for part of their life cycle (Zelo et al., 2000; Jackson et al., 2005). These species can then provide food for other key threatened or valued species (Shipman, 2001; Jackson et al., 2007). Habitat enhancement shows promise, but designs for enhancement are often species dependent, and the advantages of changing beach characteristics to accommodate target species may not be accompanied by improvements for other species.

Natural dunes provide many important ecological functions and services, including refuge areas, nesting sites, habitat for invertebrates, and corridors for migrating species. Many of the functions and services of natural dunes can be provided on artificially constructed dunes, although ecological considerations are rarely included in designs based primarily on coastal risk reduction. Even in the absence of artificial dune construction, nourishing a beach with suitable sediment will create a dune with the internal stratification, topographic variability, surface cover, and root mass of a natural dune if allowed to evolve naturally (Nordstrom, 2008). Incipient dunes will build seaward until limited by wave erosion, at which point, the most seaward portion of the dune may grow into an established foredune, which then provides protection for vegetation less well adapted to wind stress and salt spray to evolve landward of it. The creation of multiple ridges within an evolving dune field provides a variety of microhabitats, with moist slacks alternating with higher ridges. Under natural conditions, restoration of the morphology and vegetation assemblages of foredunes after storm loss can take up to 10 years (Woodhouse et al., 1977; Maun, 2004).

Most coastal risk reduction projects include artificially constructed dunes built by machines or established using fences and vegetation plantings. Artificially constructed dunes are often built as a single ridge and planted with a single species (e.g. *Ammophila* spp.) to stabilize the surface. Stabilizing the sand surface with a primary dune stabilizer can ameliorate environmental extremes and facilitate establishment of other species that are less adapted to stressful environments (Bertness and Callaway, 1994; Callaway, 1995; De Lillis et al., 2005; Martínez and Garcia-Franco, 2004). Dunes established by fences or vegetation plantings close to human facilities allow naturally functioning incipient dunes to evolve seaward, leaving a lower, moister environment between the two ridges. Dune slack environments are especially valued because they have become rare in human-altered environments (Nordstrom et al., 2012). Sand fencing is important in initial stages of dune building to create a protective dune quickly (Mendelssohn et al., 1991). However, optimal dune configuration for enhancing ecological values is one with sand fences completely buried so that they do not inhibit movement of fauna, suggesting more careful placement and fewer fences than have been used previously (Grafals-Soto, 2012; Nordstrom et al., 2012).

Optimizing all possible environmental benefits of a nourishment project would likely make project costs prohibitive, but significant increases in benefits can be achieved with minimal costs. For example, the ability of a nourished beach to provide natural habitat or enhance dune formation that could provide both risk reduction and habitat at low cost is generally underappreciated. Providing improved coastal risk reduction while also enhancing the value of natural habitat will require cooperation by local partners, especially for maintaining dunes and beaches. Counterproductive actions at the local level often include restricting the width of dunes to increase space for beach recreation, maintaining dunes at low elevations to allow for views of the sea, raking beaches to eliminate wrack (and incipient dunes) and driving on beaches. Incorporating provisions for better post-construction management of landforms and habitats by local partners would help ensure that the sediment resource evolves to a condition that provides environmental benefits as well as coastal risk reduction (Nordstrom et al., 2011). In general, beach nourishment projects have not prioritized habitat restoration as a long-term goal. Without more comprehensive project objectives, nourishment could have more adverse environmental impacts than benefits. Specific project design and management changes to maximize environmental resources can help compensate for impacts of beach nourishment.

Other Nature-Based Approaches

Natural habitats and nature-based coastal risk reduction strategies have recently attracted substantial interest due to increasing recognition of the multiple benefits they provide (e.g., Arkema et al., 2013, Duarte et al., 2013, Hettiarachchi et al., 2013). Nature-based approaches, such as restored or enhanced seagrass, salt marsh and mangrove habitats, and oyster or coral reefs, can reduce coastal erosion and wave damage and augment other structural and/or nonstructural coastal risk reduction strategies. In addition, they provide important additional ecosystem services, providing habitat that enhances commercial and recreational fisheries, improves water quality, and promotes tourism. These coastal habitats play an important role in carbon sequestration, and they have the capacity to adapt to sea-level rise (Duarte et al., 2013; Rodriguez et al., 2014). Increasingly, multiple habitats are constructed in a mosaic to maximize ecological synergies. The physical conditions under which each of these types of habitats develops differ with respect to depth, salinity, sediment grain size, tidal range, and climate conditions.

Numerous studies have documented how marine vegetation can attenuate water flow, reduce wave propagation, and stabilize sediment (e.g., Kobayashi et al., 1993; Nepf, 1999; Duarte et al., 2013; Renaud et al., 2013), thereby potentially lessening storm impacts (Gedan et al., 2011). However, the development and application of numerical models to evaluate the potential of natural systems to reduce coastal risk lag behind modeling of hard structures (Jones et al., 2012; Arkema et al., 2013). To date, studies of the performance of coastal ecosystems in hazard mitigation have generally been more qualitative in nature, although the role of these habitats in protecting landward infrastructure from storm surge and wave action is an area of intensive research (e.g., Bouma et al., 2014).

Overall Effectiveness for Reducing Flood Damage and Environmental Benefits

Healthy nearshore ecosystems represent areas of rich biological productivity that provide numerous ecological and ecosystem services. The documented effectiveness of salt marshes, oyster and coral reefs, mangroves, and seagrasses in reducing flood damage is discussed below, along with other ecological benefits.

Salt marshes. Salt marshes are dense stands of salt-tolerant plants that dominate the upper intertidal zone, in areas routinely flooded by the tides, and they are broadly distributed across the Gulf of Mexico and along the Atlantic Coast. Marshes represent some of the most threatened ecosystems in the world (Gedan and Silliman, 2009; Teal and Peterson, 2009), because flood control and navigation efforts have greatly changed or even eliminated sediment supply for these areas, limiting vertical accretion rates (Day et al., 2007). In fact, the State of Louisiana has proposed 10 large-scale sediment diversion projects in its $50 billion *Master Plan for a Sustainable Coast* to help restore coastal marshes that are steadily being lost (CPRA LA, 2012). Establishment of human-created or restored salt marshes has generally been successful (Vose and Bell, 1994; Staszak and Armitage, 2012).

Wave attenuation and shoreline stabilization are the primary coastal risk reduction benefits of salt marshes, although the quantitative effects are not fully understood. Shepard et al. (2011), in a meta-analysis of over 70 publications, noted that seven studies demonstrated that salt marsh had a significant effect on wave attenuation. Factors associated with wave attenuation included marsh width and vegetation height, stem stiffness, and density (Bouma et al., 2005; Shepard et al., 2011; Sheng et al., in press). In a meta-analysis of 15 field studies, Gedan et al. (2011) identified much greater attenuation for wind waves during low energy events than for storm surge events. Storm characteristics play an important role in the attenuation of storm surge by vegetation, with faster moving storms more effectively attenuated than slow moving storms as discussed in further detail in the section on mangroves. For example, the steady winds of Hurricane Rita overwhelmed the frictional forces of the wetland vegetation, and surge heights increased as they traveled across 25 miles (40 km) of salt marsh (Resio and Westerink, 2008). Numerous studies reported a positive effect of salt marsh vegetation on shoreline stabilization (accretion, erosion reduction, and/or positive elevation changes). No data are available on the capacity of salt marshes to reduce the extent of flooding (Shepard et al., 2011). In addition to wave attenuation and sediment stabilization, salt marshes provide essential fish habitat and improve water quality by decreasing turbidity and sequestering nutrients. Marsh edges provide an important habitat for free-swimming species, and increasing amounts of marsh edges have positive effects on shrimp abundance and survival (Minello et al., 1994; Haas et al., 2004). In the northern Gulf of Mexico, similar relationships exist for the marsh edge extent and abundance of blue crabs, spotted seatrout, and red drum.

Oyster reefs. Oyster reefs, located in both subtidal and intertidal locations, extend across a range of low to high salinities in areas with fine to sandy sediments. Fringing oyster reefs along or just offshore of vegetated shorelines may serve to dampen wave energies and increase sediment retention. Shellfish reefs (primarily oyster, but also some mussels) have been recently advocated as a nature-based approach to combat coastal erosion (Figure 3-6; Scyphers et al., 2011; Temmerman et al., 2013). Scyphers et al. (2011) reported that erosion rates decreased by 40 percent for salt marshes located behind restored oyster reefs when compared with areas without oyster reefs; however, erosion rates for both were still high (on the order of meters per year). The study was conducted during a period of high tropical storm activity, and results suggest that oyster reefs may help mitigate erosion from routine wave heights, but that oyster reefs are quickly overtopped during storms and are not effective at dealing with higher storm surge and wave heights common in tropical storms.

FIGURE 3-6 Nature-based approaches to erosion control in Mobile, Alabama. The oyster reef is seaward of the marsh edge.

As with many other nature-based approaches, the efficacy of using oyster restoration to mitigate risks associated with hurricane storm surge has not been evaluated sufficiently to offer quantitative benchmarks for coastal risk reduction (Powers and Boyer, 2014). However, depending on their geometric configuration, submerged oyster beds may function as low-crested submerged breakwaters, in which case there may be larger data sets and models to evaluate their efficacy under various physical conditions. Low-crested submerged breakwaters have been shown to decrease shore erosion rates and in some cases increase sedimentation rates in localized areas, although this is most likely associated with low- to moderate-energy events. They are generally not used to mitigate wave attack or storm surge flooding associated with high-energy events such as hurricanes. The structures can modify the geomorphology and bathymetry of adjacent areas (Hawkins et al., 2007), so a full understanding of their impact beyond the project footprint should be considered.

Like many nearshore habitats, oyster reefs are decreasing in overall areal coverage and quality (Zu Ermgassen et al., 2012). Human-made oyster reefs usually result in high densities of oysters, although sufficient vertical relief of the reef is needed for persistence (Powers et al., 2009). Oyster reefs provide essential fish habitat, nutrient sequestration, and possible water quality improvements (Grabowski and Peterson, 2007) and are capable of adapting to keep pace with sea-level rise (Rodriguez et al., 2014).

Coral reefs. Numerous studies have examined the wave attenuation properties of coral reefs. In a recent meta-analysis of 27 publications from coral reef studies in the Atlantic, Pacific and Indian Oceans, Ferrario et al. (2014) concluded that, together, the reef crest and reef flat reduced 97 percent of the wave energy that would otherwise have impacted the shoreline. This percentage energy reduction was consistent for small as well as hurricane-sized waves. Overall, the reef crest accounted for 86 percent of the reduction; half of the additional reduction occurred on the reef flat within 150 m of the crest. Therefore, even relatively narrow reefs can be effective for wave attenuation. As is the case for oyster reefs, coral reefs function similarly to low-crested submerged breakwaters and therefore physical factors such as the depth of the reef at its shallowest point and the coral composition (which will influence its roughness) are expected to have significant impact on wave attenuation. Unfortunately, these physical factors have been largely unreported in past studies. The authors concluded that wave height reductions were similar to or exceeded the benefits of constructed low-crested detached breakwaters, at a lower median cost. Although coral reefs are likely to have minimal effects for reducing storm surge (see previous discussion on oyster reefs), their value for attenuating wave energy represents a potential benefit of conserving existing coral reefs and expanding restoration efforts in storm-prone areas.

Mangroves. Mangrove vegetation dominates subtropical and tropical coastlines. Within the Gulf of Mexico, mangroves cover substantial areas of the central-southwestern shorelines of the Florida peninsula. Mangroves have historically been thought to attenuate impacts from waves and storm surge, thereby providing some risk reduction to coastal developments (Zhang et al., 2012). Mangroves also provide reduction in wind damages because of the sometimes substantial tree height (Chen et al., 2012), although Florida law allows homeowners to trim mangroves into hedges or to remove the lower tree canopy to improve visual access to the waterfront (Florida Statutes § 403.9321). In addition to wave attenuation, mangroves sequester carbon dioxide, decrease turbidity, sequester nutrients, stabilize sediments, and provide essential fish habitat (Nagelkerken et al., 2008; McCleod et al., 2011).

Field and modeling studies have demonstrated that mangrove vegetation can play a role in reducing coastal hazards. In a synthesis paper, McIvor et al. (2012) reported reductions in storm surge height of 3 to 10 inches per mile (5 to 15 cm/km) of mangrove width from a field study (Krauss et al., 2009) and 13–32 inches per mi (20–50 cm/km) from a modeling study (Zhang et al., 2012). Zhang et al. (2012) determined that surge attenuation in mangroves was nonlinear, with the greatest surge reductions (in centimeters per kilometer) occurring at the seaward edge. However, as the water hits the resistance of the vegetation, it can cause an increase in water levels in front of the mangroves, increasing the height of the storm surge there. Zhang et al. (2012) concluded that flooding would have extended 70 percent farther inland during Hurricane Wilma without the mangrove zone. The surge-reducing potential of mangroves depends on the storm characteristics. As is true with other forms of vegetation, mangroves are more effective for fast-moving storms. Zhang et al. (2012) found that storm surge from slowly-moving Category 4 and 5 hurricanes was not reduced significantly by a 10-20 mile (15- to 30-km-) wide mangrove zone.

Modeling analyses have shown that mangroves are capable of significantly attenuating short-period wind waves, reducing their height by 75-100 percent over 0.6 mi (1 km) (Mazda et al., 2006; Quartel et al., 2007). Similarly, Tanaka (2008) found that a 490–ft-wide (150-m) band of the non-mangrove tree species *Casuarina equisetifolia* did little to affect storm surge but was effective at damping short-period wind waves. Thus, mangroves and other trees appear most effective at reducing wind waves and the associated erosion these waves might cause.

Along Gulf of Mexico coastlines, mangrove habitats are generally quite narrow in developed areas (30-300 ft [10-100 m]) but can also be as wide as a 1–2 mi (2–3 km) fringe in preserves or parks. Restoration of mangroves has proven successful, but requires correct tidal elevation and physical setting for mangrove establishment. Mangrove habitat has been restored at some sites along the west coast of Florida, but a large proportion of Florida's central and southwestern coastlines, originally composed of mangrove habitat, are now protected by seawalls. Thus, further research is needed to determine the efficacy of mangroves under existing and restored conditions for reducing risk in the context of other benefits provided.

Seagrasses. Seagrasses (Figure 3-7) are the dominant forms of shallow subtidal vegetation along the Atlantic Coast and the Gulf of Mexico. Because of light limitations, their distribution is typically restricted to water depths of less than 3 m (Dennison, 1987). Numerous studies have explored how the seagrass canopies modulate water flow and currents (e.g., Fonseca et al., 1982; Gambi et al., 1990), contribute to wave attenuation, and retain and stabilize sediments in shallow coastal areas (NRC, 2007). Such sediment retention can lead to sediment accretion and reduced water turbidity.

FIGURE 3-7 An image of subtidal seagrass, *Syringodium filiforme*, from the southwest coast of Florida.

Dissipation of wave energy by seagrasses has also been proposed to play a role in reducing erosion of coastlines (Dean and Bender, 2006; Ozeren and Wren, 2010). Because seagrass canopies are relatively short (generally < 20 in [50 cm]) and flexible, substantial modification of water flow is most effective when seagrasses are found in high density and distributed over a wide area in shallow water depths (e.g., Fonseca et al., 1982; Gambi et al., 1990; Christianen et al., 2013). Because seagrasses are subtidal, frictional forces would quickly be reduced by higher water levels associated with storm surge.

Seagrasses also provide essential fish habitat and regulate nutrients in the water column. Seagrass restoration has had more limited success than salt marsh restoration efforts, and human-created sites are often of limited size (Bell et al., 2014).

Economic Costs and Return

USACE (2013f) has made some rough cost estimations for 10-year level of risk reduction related to several nature-based approaches. Total estimated first construction costs were provided for wetland restoration ($14 million/mile [$8.7 million/km]), seagrass restoration ($13 million/mile [$8.1 million/km]), and restoration of oyster reefs ($25 million/mile [$15.5 million/km]), not including operations and maintenance. Ferrario et al. (2014) estimated the costs of coral reef restoration at $2.1 million per mile ($1.3 million/km)—much less than the median reported breakwater construction cost of approximately $32 million per mile ($19.9 million/km).

Estimates of benefits are widely ranging, and little economic data are available on documented cost savings from nature-based coastal risk reduction features from prior storm events. However, Costanza et al. (2008) used a regression model on damages from 34 major U.S. hurricanes since 1980 and estimated that coastal wetlands currently provide $23.2 billion/yr in coastal storm damage reduction services. Using site-specific hurricane probabilities, calculated annual hurricane damage reduction services were ranged from $100/acre to $21,000/acre ($250/ha to $51,000/ha). In Belgium, reclaimed wetlands are being converted to marshes and floodplains for flood risk reduction at an estimated cost of $829 million, but the project is expected to be offset by flood damage reduction savings of $1.4 billion by 2100 (Broekx et al., 2011; Temmerman et al., 2013).

Healthy nearshore ecosystems also provide numerous ecosystem services, including support of commercial and recreational fisheries, nutrient regulation, shoreline stability, and other activities such as ecotourism and recreation (NRC, 2007; Temmerman et al., 2013; Powers and Boyer, 2014). These benefits could vary widely with the specific project plans.

Role of Nature Conservation

Across the United States, existing coastal habitats reduce the exposure of property to coastal storms. Without intact habitat, property and human exposures to coastal hazards would be much greater (e.g., Arkema et al., 2013). Therefore, conservation of existing coastal habitats, including intact coastal dunes and other natural coastal ecosystems (salt marshes, reefs, mangroves) has been recognized as a cost-effective risk reduction strategy with the capacity to adapt to increasing sea-level rise. Coastal development has degraded these habitats, and losses will likely continue without concerted efforts to prevent them. Federal, state, and local governments and nongovernmental organizations, such as land trusts or conservancies, work to preserve natural lands through purchase or donations of land or through conservation easements. However, in most states, land and habitat conservation has only recently been seriously considered as part of coastal risk reduction strategies. Using natural conservation areas in a comprehensive risk reduction system would need to be assessed carefully because naturally evolving coastal segments do not react in the same manner as nature-based or engineered structures. One example where this is currently playing out is Fire Island, New York, where the National Park Service is using a natural-processes strategy (Williams and Foley, 2007) to allow a new inlet created by Hurricane Sandy to remain open,[6] although some stakeholders are concerned that the breach exposes those on the mainland to greater flooding (Foderaro, 2013).

PERFORMANCE OF CONSEQUENCE REDUCTION STRATEGIES

Coastal strategies that reduce the consequences of a hazardous event involve the *location* and *design* of development. The goal of the *location* approach is to avoid or limit development in flood hazard areas before a disaster or seize opportunities for risk reduction and other community improvements after a disaster. This approach can reduce losses and protect and restore ecosystem services that reduce flooding, support biodiversity, and provide recreational activities. Tools that can be used for these purposes include land-use regulations, such as zoning; and various nonregulatory tools, such as hazard area acquisition for use as parks and greenways, purchase of repetitive-loss structures, assisting households to relocate in safer areas, and locating or relocating development-inducing critical infrastructure in nonhazard areas.

The goal of the *design* approach is to structurally strengthen buildings

[6]See http://www.nps.gov/fiis/naturescience/post-hurricane-sandy-breaches.htm.

in flood hazard areas. Where hazardous areas have advantages for development, the design approach emphasizes adjustment of building and site-design practices to reduce risk. The design approach allows economic gains to be realized, but at a cost of greater loss when disaster events exceed design standards. Tools used for the design approach include regulations that require elevation of buildings and structural strengthening. Nonregulatory tools include public education programs and low-cost loans to incentivize structural improvements as well as other subsidies. A properly conducted planning process allows communities to find the right combination of the *location* and *design* approaches.

Studies of benefits and costs of mitigation have been dominated by individual case studies of successes and failures, but this can be an obstacle to advancing proactive mitigation activities. Godschalk et al. (2009) noted that "Constituents and decision makers are often skeptical, believing that individual cases are either inapplicable to their situation or non-randomly selected to support a particular view." Two notable studies offer broader and more systematic assessments of natural hazard mitigation benefits and costs than other studies. Burby et al. (1988) estimated how floodplain land-use management programs in 10 cities influenced floodplain development trends and losses. They found that compared with the projected level of expansion in floodplain development before nonstructural floodplain management programs were enacted, floodplain development and estimated average annual flood losses a decade later were significantly lower. Floodplain development had been reduced by over 75 percent of what would have occurred without planning and management programs. Comparison of the costs and benefits of managing development showed substantial net benefits from the efforts of the 10 cities ($8.50/yr in reduced property damage for every $1 in administrative and private costs).

In response to a 1999 congressional directive, Rose et al. (2007) were funded by FEMA to conduct the most rigorous and comprehensive study of the benefits and costs of federal mitigation investments done to date. Rose et al. (2007) applied benefit-cost methods to a statistical sample of the nearly 5,500 FEMA mitigation grants to state and local governments funded between 1993 and 2003. The grants were administered under the three main federal programs that supported building design modifications and nonstructural hazard mitigation during this period: the Hazard Mitigation Grant Program, Project Impact, and the Flood Mitigation Assistance Program (see Chapter 2). Grants that were examined covered the three hazards that generate greatest losses, including floods, earthquakes, and wind. Various categories of benefits (i.e., losses that would have occurred if mitigation activity had not been implemented) were computed for each hazard and aggregated to compute an overall benefit-cost ratio by hazard and across all three hazards. Categories of benefits included, for example, reduced direct property damage, increased wetland values created due to removal of structures, reduced costs due to avoided injury or death, reduced direct and indirect losses from business disruptions, reduced nonmarket damages (e.g., historic structures), and reduced emergency response (e.g., ambulance service, fire protection).

The results for all three hazards indicate overall benefit-cost ratio for FEMA's mitigation grants of about 4:1. The benefit-cost ratio was highest at 5.0 for floods, followed by 3.9 for wind, and 1.5 for earthquakes. Flood grant benefits represent 80 percent of the total FEMA grant benefits. Rose et al. (2007) also estimate that "95% of flood benefits are attributable to avoided losses to structures and contents, and only 3% is for casualty reduction" (Rose et al., 2007). The focus of FEMA's earthquake mitigation grants has been on reduction of casualties (e.g., making schools and hospitals safe for occupants during a seismic event), but a high percentage of mitigation grants for wind hazards (hurricane and tornado) were intended to reduce the risk of business disruptions due to the vulnerability of electric utilities. Flood mitigation grants, which have emphasized reduction of property loss, had a higher benefit-cost ratio because the majority of flood grants were for residential property acquisition that had experienced repeated flooding. The authors concluded that flood grants had higher benefit-cost ratios because they are for properties with known histories that were located in the heart of mapped flood hazard areas, and recurrence of floods in a given location is more certain than for other hazards.

Participation in the Community Rating System (CRS) under the National Flood Insurance Program (NFIP) is indicative of the limited success of nonstructural approaches for risk reduction. The CRS was established in 1990 as an incentive-based voluntary program to entice better local floodplain management efforts that exceed the minimum NFIP requirements. As a CRS rating improves, local policyholder rates are reduced by up to 45 percent. On the basis of a national sample of 450 communities participating in the CRS, Brody and Highfield (2013) found that when communities adopted land-use regulations aimed at open-space protection on floodplains, which is just one of the 18 mitigation activities covered under CRS, insured flood damages under the NFIP were reduced on average by about $200,000 per year between 1999 and 2009. Despite the potential for insured loss reduction, only 5 percent of the over 21,000 NFIP-designated communities participate in the CRS, representing about 67 percent of flood insurance policies (FEMA, 2012). A study of 71 communities in Florida and North Carolina indicates that CRS incentives are too small and inconsequential to motivate communities to adopt and implement land-use policies in their hazard mitigation plans under the CRS program (Berke et al., 2014). Instead, this study found that state mitigation policy has a stronger influence on inclusion of land-use policies in hazard mitigation plans aimed at avoiding flood hazard areas.

Studies consistently indicate that where plans aimed at hazard mitigation have been adopted, they foster robust local hazard mitigation programs and a reduction in property damage in natural disasters. Evidence shows that applying measurable indicators of the strength of hazard mitigation plans led to stronger local programs and thus lower losses to property. In studies of California earthquakes (including the 1994 Northridge earthquake), the probability of significant damage to a building was lower in areas that had robust mitigation plans (Olshansky, 2001; Nelson and French, 2002). Studies in Australia, New Zealand, and the United States have documented a number of benefits that follow when local governments have developed mitigation plans (May et al., 1996; Ericksen et al., 2004; Berke et al., 2006). These benefits include increased knowledge about hazards among local decision makers and greater linkage with other local issues in ways that helps prioritize mitigation efforts. For example, acquisition of severe repetitive loss structures along greenway corridors that straddle floodprone coastal waterways offers multiple co-benefits, including reduction in future damages, increased recreational access for the general public, and improved opportunities for physical activities that yield public health benefits (Younger, 2008), thereby expanding stakeholder support for the mitigation project and the likelihood of implementation (Ostrom, 2010).

REGIONAL AND NATIONAL IMPLICATIONS OF SAND MANAGEMENT

Coastal risk reduction involves significant cost, whether by nature-based strategies such as beach nourishment and wetlands restoration, or hard structures such as levees, seawalls, and storm surge barriers. In addition to the costs of planning, design, construction, and maintenance, resources such as sand, mud, and other materials are needed for construction or restoration. Sediment, particularly sand, is a key resource for coastal risk reduction. Beach nourishment projects often require hundreds of thousands to millions of cubic yards of sand. The sources for this material are usually offshore sand deposits but can be onshore sources as well (NRC, 1995). The offshore deposits are due in part to glacial processes and to relic beaches left on the continental shelf as the sea level rose almost 400 ft (120 m) over the last 20,000 years, moving the shoreline landward (Williams et al., 2012; Figure 3-8). Other offshore sand deposits nearer the shoreline are due to modern-day coastal and estuarine processes, such as shoals created by tidal flows at inlets and bays.

Periodic applications of sand are necessary to maintain beach nourishment projects, which can lead to the removal of large total volumes of offshore sand. The environmental impacts of dredging were discussed earlier in the chapter (see Beach Nourishment and Dune Building). Nearby offshore sources can become depleted—for example, this is a current prob-

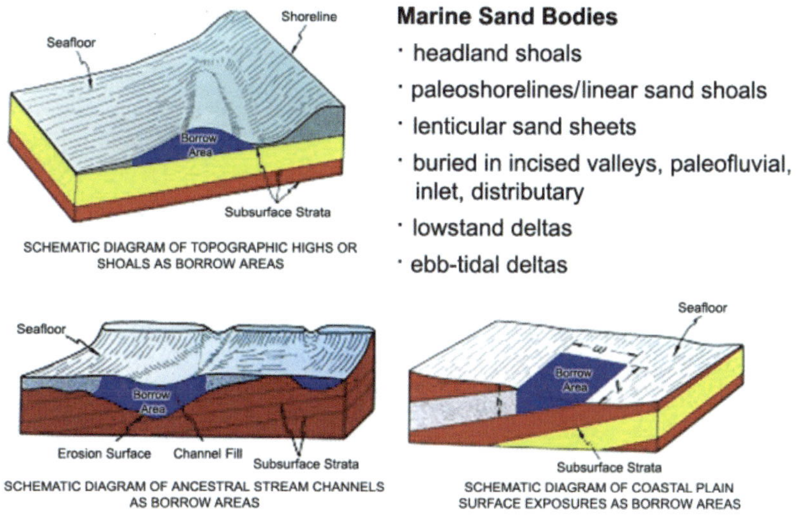

FIGURE 3-8 Various types of offshore sand bodies that are potential sources of nourishment sand.
SOURCE: http://woodshole.er.usgs.gov/project-pages/aggregates/overview.htm.

lem for offshore sand sources in Miami-Dade County, Florida, although the southeast Florida region as a whole has an excess of sand (Ousley et al., 2012). In situations where offshore sources become depleted, additional sources will have to be found. Although federal and state agencies have documented offshore deposits suitable for beach nourishment, not all are conveniently located to meet project needs within available funding. While overall U.S. sand and gravel resources are plentiful, the availability of high-quality beach sand near shorelines could become scarce, which could increase the costs of future beach nourishment projects.

In most instances, development along the landward edge prevents beaches from naturally migrating landward in response to rising sea levels. Instead, most beaches will have to be at a higher elevation, which will increase the demand for sand. Leatherman (1989) examined the amount of sand needed to maintain all U.S. recreational beaches and estimated that 4.3 billion yd^3 (3.3 billion m^3) of material would be required to deal with 3.3 ft (1 m) of sea-level rise.

Although Williams (1986) estimated sand and gravel resources within the U.S. exclusive economic zone at more than 1,600 billion yd^3 (1,200 billion m^3) in water less than about 200-ft (60-m) deep (NRC, 1995), much of this material lies outside the jurisdictions of the various coastal states and would require negotiations with the federal government. Costs to extract this federal resource are likely to be more expensive due to factors such as water depth and distance from the site.

The Bureau of Ocean Energy Management (previously the Minerals Management Service [MMS]) manages sediment resources in federal waters, and has worked with New York, New Jersey, Delaware, Maryland, Virginia, North Carolina, Florida, Alabama, Louisiana, and Texas to identify and characterize offshore sand deposits that could be available for nourishment sand. For example, from 1992 to 1995 the MMS funded the Maryland Geological Survey to examine offshore sand deposits near the state's barrier island beaches.[7] They identified a number of long linear sandy shoals containing 330 million yd^3 (250 million m^3) of beach-quality sand within 24 miles (39 km) of the Maryland coast, in waters less than 50 ft (15 m) deep. The U.S. Geological Survey, in conjunction with several other institutions, has built usSEABED,[8] a database of offshore sediment data for the entire United States that can be used to map possible sand regions for beach nourishment projects. The USACE also has a Regional Sediment Management Program to manage sediment across regions with multiple projects.[9]

Nearshore dredging from inlet ebb and flood shoals, although difficult because of wave exposure, is a valuable source of beach-quality sand. The USACE has historically disposed of this sand offshore as the low-cost alternative, although there have been some instances where beach renourishment has been carried out using sand dredged from the navigational channel (e.g., Perdido Key, Florida; Browder and Dean, 2000; Wang et al., 2013). Terminal structures that block longshore transport, whether natural (such as spits and capes) or artificial (such as jetties), can serve as source areas for beach nourishment.

Beach-quality sand can be retained regionally through bypassing and backpassing projects. "Bypassing" transfers sediment past obstructions to longshore transport, such as at inlets maintained by jetties or dredging. "Backpassing" transfers sediment from accreting downdrift areas back to the updrift locations, and is often conducted in small amounts and with equipment available at the local level (Mauriello, 1991). On a larger scale, backpassing can occur where sediment deposited at inlets or at the depositional end of spits is returned to updrift beaches (Cialone and Stauble, 1998). The use of sediment that is already in the regional coastal system ensures compatibility with beach material and avoids issues associated with dredging and borrow pits.

[7]See http://www.mgs.md.gov/coastal/osr/mosr1.html.

[8]See http://walrus.wr.usgs.gov/usseabed/.

[9]See http://rsm.usace.army.mil/.

CONCLUSIONS AND RECOMMENDATIONS

Within the past few decades, as adverse environmental impacts of hard structures became clear, most USACE coastal storm damage reduction projects have emphasized beach nourishment and dune building. This chapter reviewed what is known about the proven performance, costs, and benefits of hard structures and nature-based strategies to reduce the hazards (e.g., flooding, wave attack) associated with coastal storms and nonstructural and building design measures to reduce the consequences of coastal hazards. Determination of the optimal coastal risk reduction will be site specific and depend on an analysis of long-term costs, benefits, and environmental impacts, and may involve multiple approaches implemented together.

Beach nourishment and dune-building projects for coastal risk reduction can be designed to provide increased ecological value. Beachfill projects provide some level of risk reduction for coastal infrastructure from erosion, flooding, and wave attack and may reduce the likelihood of forming new inlets. Beach nourishment and dune building do not, however, address back-bay flooding, which may be better addressed by structural measures on the bayside. The short-term environmental impacts of nourishment projects on biological communities is significant, and long-term cumulative ecological implications remain unknown because of the difficulty and cost of mounting large-scale monitoring projects and the limited time frame of existing studies. Coastal systems can be managed for multiple uses and benefits, although some compromises may be necessary to optimize benefits across a range of objectives. Improvements for ecological benefits of beach nourishment and dune construction would involve different design specifications that are unlikely to greatly increase construction costs, although they may require alternative approaches to post-construction beach and dune management.

Sediment management should be viewed on a regional basis, rather than on a project-by-project basis. Federal and state agencies have documented plentiful offshore sand deposits for beach nourishment but not all are of optimal quality or conveniently located to project needs, which could increase costs. Coastal projects can minimize sediment losses by retaining dredge material or emphasizing reuse, as in sand backpassing or bypassing operations. Use of a sediment source that is compatible with a beachfill project site also decreases ecosystem recovery time and enhances habitat value in the nourished area.

Conservation or restoration of ecosystem features such as salt marshes, mangroves, coral reefs, and oyster reefs provides substantial ecological benefits and some level of risk reduction against coastal storms, but the risk reduction benefits remain poorly quantified. Coastal habitats provide numerous ecosystem services, including carbon sequestration, improved water quality, and essential habitat for commercial and recreational fisheries. Much is known about the capacity of nature-based features to reduce coastal erosion from smaller storms, but additional research is needed to better understand and quantify the effects of natural features (other than beaches and dunes) on storm surge, wave energy, and floodwater inundation. In general, the level of risk reduction provided by oyster reefs and seagrasses appears much lower than that provided by constructed dunes and hard structures, and most of the benefits are associated with reductions in wave energy during low- to moderate-energy events. Research has documented reductions in peak water levels from salt marshes and mangroves, but certain storm conditions and large expanses of habitat are needed for these to be most effective. Thus, many of these nature-based alternatives can only be used for coastal risk reduction at locations that have sufficient space between developed areas and the coastline. Additional quantitative modeling and field observation are needed to better understand and quantify the efficacy of nature-based approaches for coastal risk reduction.

Hard structures are likely to become increasingly important to reduce coastal risk in densely populated urban areas. Many large coastal cities lack the space necessary to take advantage of nature-based risk reduction approaches alone and will instead need additional hard structures to substantially reduce coastal hazards. Adverse environmental impacts commonly accompany the construction of hard structures, although modified designs are possible to reduce these effects. Coupling nature-based approaches with hard structures to buffer the structures against wave attack provides an effective coastal risk reduction strategy if space allows.

Strategies that reduce the consequences of coastal storms, such as hazard zoning, building elevation, land purchase, and setbacks, have high documented benefit-cost ratios, but they are given less attention by the federal government and are viewed as difficult to implement by states. Studies have reported benefit-to-cost ratios between 5:1 and 8:1 for nonstructural and design strategies that reduce the consequence of flooding, but between 2004 and 2012, federal funds for mitigation were only about 5 percent of disaster relief funds. Those nonstructural and design strategies that are commonly implemented, such as public information campaigns and elevation of in-situ development tend to avoid property rights issues, do not threaten economic interests, and do not generate political opposition.

4

Principles for Guiding the Nation's Future Investments in Coastal Risk Reduction

Investments in coastal risk reduction measures generate significant benefits to society by reducing damage to buildings and infrastructure from coastal storms and potentially saving lives. Investments in coastal risk reduction also involve significant costs (e.g., construction and maintenance, meeting upgraded building codes, and loss of benefits from restrictions on development in vulnerable areas). Investments in coastal risk reduction may also have impacts on coastal ecosystems, which may generate additional costs (or benefits) through changes in the provision of ecosystem services (see Chapter 3). Investment in coastal risk reduction can take many forms including strategies to reduce the probability of a hazard (e.g., seawalls, surge barriers, dune construction, and marsh restoration), and strategies to reduce the consequences of a storm event (e.g., building codes, zoning requirements, and strategic retreat from vulnerable coastal areas). Different strategies for coastal risk reduction will differ in terms of their risk reduction benefits, costs, and ecosystem impacts. A key question facing society is determining when investments in coastal risk reduction are justified, and if justified, what form they should take. These decisions can be made at national, regional, state, or local levels and involve input from a broad array of stakeholders (those that have an interest in or are affected by decisions regarding coastal risk).

The committee was tasked to address the following questions: How might risk-related principles contribute to the development of design standards for coastal risk reduction projects? What general principles might be used to guide future investments in U.S. coastal risk reduction? (See Chapter 1.) This chapter describes and compares two approaches to determining what investments in coastal risk reduction are worthwhile: (1) a risk-standard approach (sometimes called a "level of protection" approach) and (2) a benefit-cost approach. The *risk-standard approach* recommends investments in coastal risk reduction measures to achieve an acceptable level of risk reduction, such as reducing the threat of loss of life (e.g., 1 in 1,000 chance annually of more than 50 deaths in a single event) or the probability of severe flooding (e.g., 1 in 200 chance annually of overtopping a levee system). Thus, the risk-standard approach considers a specific consequence and designs cost-effective strategies to alter the probability of that consequence occurring. The *benefit-cost approach* recommends investment in coastal risk reduction when the benefits of the investment exceed the costs, considering both probability and consequences along a continuum of possible events. Thus, the level of risk reduction provided by projects under a benefit-cost approach is not predetermined but would vary based on the costs and benefits provided.

Differences between the risk-standard and benefit-cost approaches are illustrated in Figure 4-1, which presents a hypothetical plot of risk versus net benefits (benefits minus costs) from risk reduction investments. The benefit-cost approach would advocate investment in risk reduction to the point that maximizes net benefits (Point B). A risk-standard approach requires that investments be made so that risks are reduced to (or below) the acceptable risk (Point C). As drawn, the level of risk reduction that maximizes net benefits, Point B, does not satisfy the acceptable-risk standard. Acceptable-risk standards may thus be considered as a way to constrain benefit-cost outcomes. However, it is also possible that the level of risk reduction that maximizes net benefits is actually well beyond the level needed to meet the acceptable-risk standard (if plotted, Point B would lie to the left of Point C). In urban areas, providing risk reduction measures beyond the acceptable-risk standard could be a wise investment given the large value of property being protected and the potential savings in human lives. For example, even if an acceptable-risk standard is determined to be a 1 percent annual-chance (100-year) event, net benefits in coastal cities might be maximized by providing risk reduction measures designed for a 0.2 or 0.1 percent annual-chance (500-year or 1,000-year) event. For purposes of clear exposition in the sections that follow, these two approaches are treated as separate and distinct. In reality, however, blending elements of each into a hybrid approach may be desirable, as is discussed in more detail in the final section of the chapter.

A RISK-STANDARD APPROACH

A risk-standard approach establishes an acceptable risk and makes investments so that risks are reduced below this level. For example, an agency

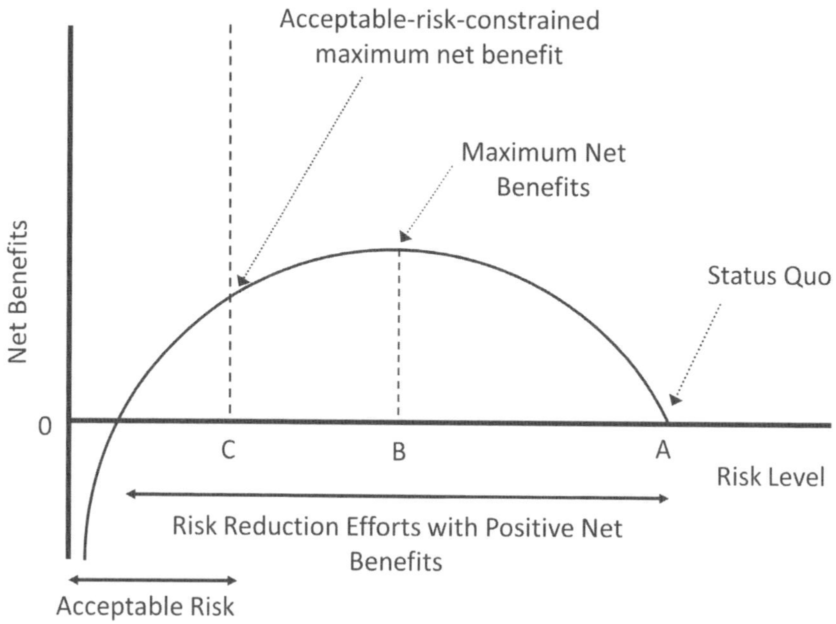

FIGURE 4-1 Illustration of the risk-standard and benefit-cost approaches. The horizontal axis shows the levels of risk while the vertical axis shows net benefits. Starting from Point A, the status quo, investments in risk reduction initially have positive net benefits as the value of risk reduction exceeds the costs of investment. As risk levels decline toward zero, the costs of making further investments increase faster than the value of risk reduction and net benefits of further risk reduction begin to fall, eventually driving net benefits negative. The optimal risk reduction according to the benefit-cost approach falls at Point B and results in a higher level of risk (in this scenario) than the risk-standard approach, which is constrained by acceptable risk.

may design measures to eliminate or substantially reduce risk for events more frequent than a congressionally mandated level of risk reduction. This can be accomplished either by reducing the probability of the hazard (e.g., by building appropriately sized levees or dunes to substantially reduce risks up to a certain magnitude event) or by eliminating the consequences (e.g., by abandoning an impacted area or by elevating structures above the flood depth). Although risk standards do not consider costs explicitly, the risk-standard approach often implicitly considers costs, and decision makers in collaboration with stakeholders may choose to adjust the risk standard to allow for greater risk when significant costs are involved.

Applying a risk-standard approach requires two things. The first is a risk assessment that analyzes the probabilities and consequences of coastal hazards (see Box 1-2) and evaluates how investments could reduce either, thereby reducing risk. The second requirement for applying an acceptable-risk standard is getting agreement on what is acceptable versus unacceptable risk, which is discussed in more detail in the next section.

Acceptable Risks

Rarely, if ever, is it possible to reduce risk to zero. If some level of risk is unavoidable, or could only be avoided at extreme cost, what is a low-enough level of risk that is satisfactory to stakeholders? How to define acceptable risk in this sense is a challenge. The acceptability of risk is not a technical question—it is a question of politics, economics, values, and ethics.

Standards for acceptable risks from a societal viewpoint have been developed in a number of countries including the United States, the United Kingdom, the Netherlands, Australia, and New Zealand. A prominent example of the application of acceptable-risk standards is in dam safety. Catastrophic failure of a dam can result in fatalities from flooding downstream of the dam. Dam safety programs model risks as a function of probability (in terms of the frequency of dam failures per year) and consequence (in terms of the number of fatalities per event). This approach is colloquially known as an FN curve because it plots the annual frequency (F) on one axis and the number fatalities (N) on the other (Figure 4-2). In the United States, the U.S. Bureau of Reclamation (USBR) led the way in using the FN curve to define the potential fatalities if certain specified dam failures occurred. This approach became an integral part of USBR practice in appraising dam safety risks and in making decisions on remedial actions to reduce safety risks at particular facilities (USBR, 2003). The U.S. Army Corps of Engineers (USACE) has recently adopted similar FN curve criterion for its dam safety program.

Acceptable-risk standards based on protection against loss of life are useful in dam safety but have limitations when considered for coastal risk reduction projects. The first limitation is that such acceptable-risk standards principally relate to catastrophes involving loss of life. Although they can be applied to financial costs, environmental impacts, or other non-loss-of-life consequences, to date this has been less common. Second, the residual risk associated with coastal risk reduction projects in the United States is much greater than residual risk permitted in modern dam safety criteria, in part because dams represent an engineered hazard, whereas coastal risk reduction projects are designed to protect against natural hazards. This

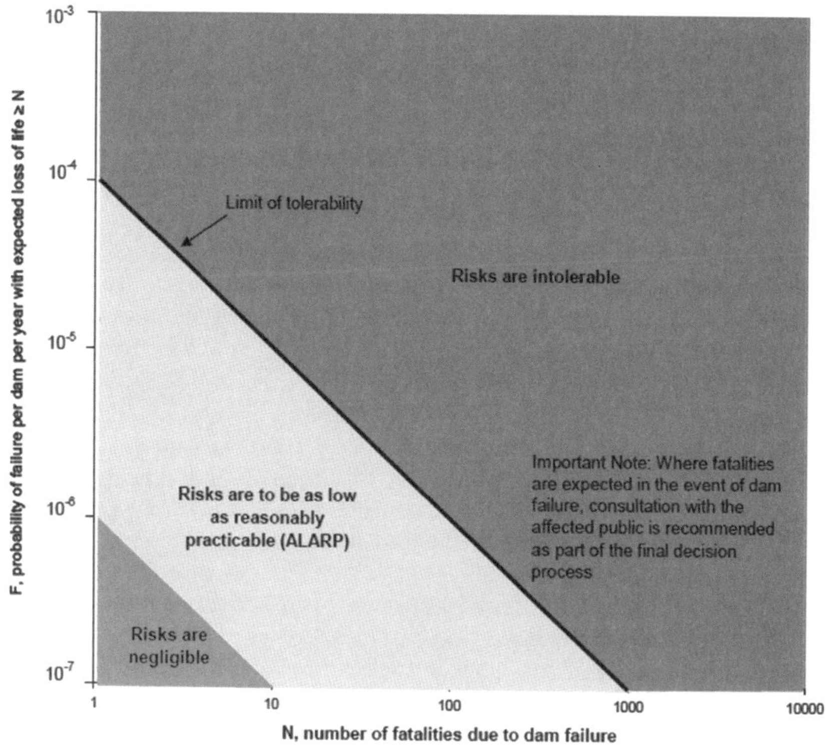

FIGURE 4-2 Dam safety societal risk requirements for new dams and major augmentations showing unacceptable (intolerable) risks (upper right) and acceptable risks that may be mitigated if practicable or are negligible (lower left). Risk is a function of the probability of an event (frequency, F) and its consequence (number of fatalities, N).
SOURCE: Reprinted, with permission, from NSW-DSC (2006).

inconsistency would require resolution before the criteria could reasonably be applied to the evaluation of coastal risk reduction projects.

Acceptable risk in more general contexts beyond evaluating risk to loss of life has proven hard to define. For example, The World Health Organization in addressing standards for water quality provides a number of different concepts that could be used in defining acceptable risk (Fewtrell and Bartram, 2001):

> A risk is acceptable when: it falls below an arbitrary defined probability; it falls below some level that is already tolerated; it falls below an arbitrary

> defined attributable fraction of total disease burden in the community; the cost of reducing the risk would exceed the costs saved; the cost of reducing the risk would exceed the costs saved when the "costs of suffering" are also factored in; the opportunity costs would be better spent on other, more pressing, public health problems; public health professionals say it is acceptable; the general public say it is acceptable (or more likely, do not say it is not); politicians say it is acceptable.

The concept of acceptable risk is defined by the United Nations (UNISDR, 2009) as "the level of potential losses that a society or community considers acceptable given existing social, economic, political, cultural, technical and environmental conditions." The United Nations further states:

> In engineering terms, acceptable risk is also used to assess and define structural and non-structural measures that are needed in order to reduce possible harm to people, property, services and systems to a chosen tolerated level, according to codes or "accepted practice"' which are based on known probabilities of hazards and other factors.

Acceptable-risk standards can be set based on some objective quantifiable risk standard. For example, acceptable-risk levels that involve personal injury and death can find guidance in such concepts as the quality-adjusted life-years (Hunter and Fewtrell, 2001). Acceptable risk standards have also been applied in industry. Starting in the 1960s, the United Kingdom's Health and Safety Executive developed criteria for "broadly acceptable" risks that appear to reflect ambient industrial risks that society willingly accepts, and "tolerable" risks that appear to reflect the highest industrial risks that society accepts if a corresponding benefit is derived (HSE, 2001; Jonkman et al., 2008, 2011). The above acceptable-risk standards have been developed with varying degrees of stakeholder involvement.

One issue with the concept of an acceptable-risk standard is that it singularly highlights one particular level of risk. In reality, less risk is preferable to more risk so that all reductions in risk have value and not just those that reduce risk to the acceptable-risk standard. The benefit-cost approach to risk reduction (discussed later in the chapter) allows for positive marginal benefits of risk reduction over the entire range of risk reduction.

Risk Perception and Setting Acceptable Risk

Setting the level of acceptable risk for any hazard is not a purely scientific or engineering matter but rather involves a societal value judgment. In a democratic society, the involvement of the public is essential for setting acceptable-risk standards. Views on what is acceptable by the public can then be combined with technical analysis to determine what investments are necessary to meet the threshold levels of acceptable risk. The National Research Council (NRC, 1996) recommended adopting an analytic-deliberative approach with technical experts, public officials, and affected parties taking part in all steps of problem formulation, assessment, and policy recommendation. Hunter and Fewtrell (2001) suggest that the process for assessing acceptable risk should bring together experts with technical knowledge with the affected public. The experts would quantify the impacts of alternatives and present recommendations to be reviewed by all stakeholders. Critics of technocratic or expert judgment point to evidence that individuals and experts do not always agree on what risks are most important to address or to what degree risks should be reduced. Often, rankings of risk by individuals do not align with rankings of risks by experts based on the best evidence of relative risks (Slovic, 2000). If acceptable risk is a level of risk that a "society or community considers acceptable," then it is a societal or community view of risk rather than expert or technocratic views of risk that are essential.

Social psychologists have provided ample evidence that risk perception varies by individual and circumstance and that "objective" statistics relied on by experts to assess risks only tell part of the story. The U.S. Environmental Protection Agency (EPA, 1988) concluded that "No fixed level of [individual] risk could be identified as acceptable in all cases and under all regulatory programs." Many factors influence the public's perception of risk: the voluntary or involuntary nature of the risk, the potential for catastrophe, the degree of familiarity, scientific uncertainty, the sense of dread, inequitable distribution of risks and benefits, and potentially irreversible effects among other things (Allen et al., 1992; Slovic, 2000; NRC, 2012b). For example, acceptable risk standards vary based on the perceived degree of voluntariness (Vrijling et al., 1995, 1998). The National Earthquake Hazard Reduction Program (Building Seismic Safety Council, 1995) established an "acceptable" probability of death for continuous occupancy in an engineered building in a high-seismicity area at 1 in 1 million (1×10^{-6}) per person per year (Porter, 2002). This risk can be compared to other risks such as 2×10^{-7} fatal accidents per departure for commercial airline travel[1] or the annual risk of dying in a motor vehicle accident of 1.1×10^{-4}.[2] The much higher accepted risk for motor vehicles noted above is often attributed to the more voluntary, and more routine, nature of the activity and the risk. The public appears willing to accept a risk up to 1,000 times greater for a voluntary risk compared with an involuntary risk (Starr, 1969).

Setting acceptable coastal risk standards would, therefore, be challeng-

[1]See http://www.ntsb.gov/data/aviation_statistics.html.
[2]See http://www.highwaysafety.org.

ing and would require extensive stakeholder engagement, including members of the public, private interests, and relevant agencies at local, state, and federal levels. All parties would need to collectively consider the risks, societal perceptions of these risks, and the willingness of all parties to pay to reduce those risks.

The 1 Percent Chance (100-Year) Level of Risk Reduction

The 1 percent annual-chance event (see Box 1-3) is a commonly applied level of risk reduction in many inland flood control projects and some coastal risk reduction projects. Until the mid-1970s, Congress supported relatively high levels of risk reduction (e.g., the Standard Project Hurricane [see Box 1-2]). However, the establishment of the 1 percent chance (100-year) event to define the special flood hazard area for the National Flood Insurance Program altered the perception of flood risk. When mandatory flood insurance purchase requirements were waived beginning in the mid-1970s for properties located behind structures designed for a 1 percent chance flood, the 1 percent level of flood risk reduction became a de facto standard for many communities (NRC, 2013). The USACE no longer uses the 1 percent chance event as a standard basis of design for inland or coastal projects (see Chapter 2), although local sponsors often request and fund the additional costs for this level of risk reduction to eliminate flood insurance requirements for residents in flood hazard areas. On some projects, such as the USACE Hurricane Storm Damage Risk Reduction System of New Orleans, this criterion was specified in congressional legislation. Thus, the 1 percent chance event is usually selected as the basis of risk reduction efforts without an explicit calculation of the benefits and costs. Is a 1 percent annual chance of flooding a better choice than say a 2 percent chance or a 0.2 percent chance for coastal risk reduction? Also, why provide the same level of reduction in the probability of flooding to both a densely populated urban area with large immovable structures and a sparsely populated rural area with little in harm's way? Surely what is at risk should also matter in designing coastal risk reduction investments.

A BENEFIT-COST APPROACH

In analyzing whether a given coastal risk reduction investment is worthwhile, the benefit-cost approach assembles evidence on the likely benefits and costs of the investment relative to the status quo (NRC, 2004b). For example, if investing in coastal risk reduction reduces the likelihood of flood damage to properties, the benefits of the investment for these properties can be found by evaluating the reduction in damages from storm events of various magnitudes, and multiplying this by the probability of occurrence of storms of these magnitudes. Consistent with the definition of risk, analysis of the benefits of coastal risk reduction measures requires an assessment of both the probability of a hazard occurring and the consequence (change in benefits) if it occurs.

Unlike the risk-standard approach, the benefit-cost approach measures the value of risk reduction benefits in monetary terms. Measuring risk reduction in monetary terms is necessary to be able to compare benefits with costs in a common monetary metric. By using a common monetary metric, the benefit-cost approach also allows for incorporation of other costs or benefits associated with coastal risk reduction strategies, such as the value of reduced damages to property, loss of life, or business interruptions, as well as changes in the value of ecosystem services. It can be difficult to quantitatively include other benefits besides risk reduction in the risk-standard approach.

Strict adherence to benefit-cost analysis would recommend funding only those investments where benefits exceed costs. The benefit-cost approach can also be cast in terms of a return-on-investment (ROI) approach, which compares the ratio of the benefits to the cost and recommends investment in coastal risk reduction measures when the benefit-to-cost ratio exceeds a certain threshold. Setting the ROI threshold equal to 1 generates the same outcome as the benefit-cost approach. When faced with a binding budget constraint so that not all projects with positive net benefits can be funded, ROI can be used to set investment priorities. Projects can be ranked by ROI and, starting with the highest ROI, funding can be allocated to the next highest ROI-ranked project until the budget constraint is met.

In discussing environmental, health, and safety regulations, Arrow et al. (1996) state that "benefit-cost analysis can help illuminate the trade-offs involved in making different kinds of social investments. In this regard, it seems almost irresponsible to not conduct such analyses, because they can inform decisions about how scarce resources can be put to the greatest social good." However, benefit-cost analysis is not the only useful information that should be considered by decision makers. Arrow et al. (1996) also state that benefit-cost analysis is "neither necessary nor sufficient for designing sensible public policy. If properly done, it can be very helpful to agencies in the decision-making process."

Measuring the Benefits of Coastal Risk Reduction Investments

Although the basic logic of benefit-cost analysis is quite straightforward, there are a number of difficult issues in applying benefit-cost analysis to investments in coastal risk reduction. One of the most difficult issues involves accurately measuring the benefits of investments in coastal risk reduction in monetary terms, as needed in benefit-cost analysis.

Investment in coastal risk reduction can potentially provide a wide array of benefits such as reduced damages to property and infrastructure, reduction in injury or loss of life, reduced social disruption for coastal communities, and improvement in an array of ecosystem services. Some of these benefits are relatively easy to measure in terms of monetary value, at least in principle. Damage to property and infrastructure can rely on information about loss of property value from storm events or flooding. Other benefits are much more difficult to measure in monetary terms. Valuing reduced disaster-related fatalities, increased socioeconomic stability for coastal communities, or restored ecological functions in monetary terms seems like a tall order. However, over the past half-century economists have developed an array of methods for estimating "nonmarket" value associated with environmental and social benefits that are often thought of as being difficult to impossible to value in monetary terms (Freeman, 2003). For example, economists have developed estimates of the value of clean air, clean water, or access to natural areas by looking at the premium in property values for otherwise similar properties located in areas with different environmental amenities (Harrison and Rubinfeld, 1978; Smith and Huang, 1995).

Even with advances in nonmarket valuation methods and applications, there remain large gaps in the ability to accurately measure benefits. Attempts to measure certain environmental or social benefits in monetary terms remain controversial. When the *Exxon Valdez* ran aground and spilled oil in Prince William Sound in Alaska, various parties sued Exxon for damages from the oil spill. Courts had to wrestle with questions about how much should Exxon pay to account for damages to the environment and various affected communities. These cases took well over a decade to litigate and spawned heated debate about the legitimacy of various methods to estimate nonmarket values associated with environmental degradation of the Sound (see, e.g., the debate over contingent valuation between Hanemann [1994] and Diamond and Hausman [1994]).

More fundamentally, some critics of benefit-cost approaches claim it is wrong-headed to try to boil down all values into monetary terms (Kelman, 1981; Sagoff, 1988). For example, how can biodiversity or spiritual and cultural values be evaluated in monetary terms? Even trying to do so might change how people think about these values and thereby distort these values. According to these critics, economic accounting should be restricted to market goods and services, and there should be separate consideration of other social and environmental values.

Another critique of benefit-cost analysis revolves around issues of equity and fairness (Ackerman and Heinzerling, 2002). Critics of benefit-cost analysis point out that the rich often get greater weight in benefit-cost analysis simply because they have more money. For example, consider a coastal risk reduction project for a community with 10 homes each worth $1 million versus another coastal risk reduction project for 50 homes each worth $100,000. The first project reduces the risk to property worth

$10 million while the second reduces risk to property worth $5 million. If both projects cost the same amount of money, benefit-cost calculations would favor doing the first project over the second. However, many observers would favor the second project over the first, in part because it affects more people and the people affected may have less ability to cope with loss.

USACE Benefit-Cost Analysis in Coastal Risk Reduction

Benefit-cost analysis has been used widely to evaluate government programs including investments in water projects (Howe, 1971; Brouwer and Pearce, 2005) and investments in environmental improvement under laws such as the Clean Air Act (EPA, 2011). The USACE has a long history of doing benefit-cost analysis dating back to the Rivers and Harbors Act of 1902 (and possibly earlier; Hammond, 1966). By the 1930s, benefit-cost analysis was well established as accepted practice, and the Flood Control Act of 1936 required that benefits exceed costs for USACE project approval. The U.S. Inter-Agency River Basin Committee, Subcommittee on Benefits and Costs (1950) produced a report known as the "Green Book" that attempted to standardize economic evaluation procedures required under the 1936 Act. Although the Green Book was never formally adopted, the Bureau of the Budget built upon the report in the development of Circular A-47 (Executive Office of the President, 1952), which established rigorous standards for evaluating federal water projects.

The 1965 Water Resources Planning Act and the *Principles and Standards* (WRC, 1973) that resulted from that legislation further shaped benefit-cost analyses for federal water resources project planning. The *Principles and Standards* required that four accounts be used for evaluating federal water projects—national economic development (NED), regional economic development, environmental quality, and social well-being—accounts that continue to be used in USACE planning today. Environmental quality and NED were originally established as coequal objectives, but this made large, structural engineering projects hard to justify (NRC, 2004b). In 1983, the *Principles and Standards* were repealed and replaced by the *Principles and Guidelines* (WRC, 1983; see also Chapter 2), which established a single objective for federal water resources projects "to contribute to national economic development consistent with protecting the Nation's environment."

At the time of the writing of the *Principles and Guidelines*, it was felt that other nonmarket environmental and social benefits could not be accurately evaluated in monetary terms, and these factors continued to be considered in separate accounts for Environmental Quality and Other Social Effects. NED—the increase in the value of marketed goods and services plus project-related recreation benefits,[3] minus construction, operations, and maintenance costs (USACE, 2011a)—became the most important decision criterion in the USACE planning framework. Aside from major adverse environmental impacts, environmental and social effects no longer significantly influenced water resources decisions (see also Chapter 2; a more detailed history is provided in NRC [1999, 2004b]). Although these policies remain in effect, there is ongoing vigorous debate about the principles and procedures governing the use of benefit-cost analyses in federal water resources projects, and as discussed in Chapter 2, Congress in WRDA 2007 directed the administration to revise the *Principles and Guidelines*.

What Should Count as a Benefit or a Cost?

Questions about whether all of the benefits or costs of investments in coastal risk reduction can be accurately measured in monetary terms raises the issue of whether benefit-cost analysis should attempt to be inclusive of all benefits and costs or whether there should be multiple accounts that are evaluated in different currencies that are not directly comparable (Polasky and Binder, 2012). A fully inclusive approach where everything is measured in a single monetary metric is appealing because it makes it easy for decision makers to compare outcomes and results in a simple and transparent decision rule based on net benefits. But if it is not possible to accurately assess all values in a common metric, then net benefits could systematically under- or overweight some benefits and generate biased decisions. In cases where there are important social, cultural, or ecosystem benefits that are difficult to quantify or monetize, it may be preferable to keep multiple accounts and set standards for acceptable outcomes for each account, or to use some form of multicriteria decision analysis (Keeney and Raiffa, 1993).

By creating separate accounts for environmental and social benefits but focusing on NED as the primary account, the *Principles and Guidelines* favor projects that score well in terms of value of marketed goods and services while giving inadequate weight to nonmarketed environmental and social benefits. The first part of the congressionally mandated revisions to the *Principles and Guidelines*—the 2013 *Principles and Requirements*, developed by the White House Council on Environmental Quality (CEQ, 2013)—give expanded consideration to environmental and social benefits rather than giving primacy to economic development (NED).

The *Principles and Requirements* summarized the limitations of the earlier approach:

[3]Under current guidance, recreation benefits may not exceed 50 percent of the overall project benefits.

> Heretofore, Federal investments in water resources have been mostly based on economic performance assessments which largely focus on maximizing net economic development gains and typically involve an unduly narrow benefit-cost comparison of the monetized effects. Non-monetized and unquantified effects are often included in the overall analysis process, but are not necessarily weighted as heavily or considered key drivers in the final decision making process. As a result, decision making processes are, at this point in time, unnecessarily biased towards those economic effects that are generally more easily quantified and monetized. A narrow focus on monetized or monetizable effects is no longer reflective of our national needs, and from this point forward, both quantified and unquantified information will form the basis for evaluating and comparing potential Federal investments in water resources to the Federal Objective. This more integrated approach will allow decision makers to view a full range of effects of alternative actions and lead to more socially beneficial investments.

The *Principles and Requirements* (CEQ, 2013) emphasizes including all benefits and costs in a common framework where feasible, via an ecosystem services approach:

> The ecosystems services approach is a way to organize all the potential effects of an action (economic, environmental and social) within a framework that explicitly recognizes their interconnected nature. The services considered under this approach include those flowing directly from the environment and those provided by human actions. Services and effects of potential interest in water resource evaluations could include, but are not limited to: water quality; nutrient regulation; mitigation of floods and droughts; water supply; aquatic and riparian habitat; maintenance of biodiversity; carbon storage; food and agricultural products; raw materials; transportation; public safety; power generation; recreation; aesthetics; and educational and cultural values. Changes in ecosystem services are measured monetarily and non-monetarily, and include quantified and unquantified effects. Existing techniques, including traditional benefit costs analyses, are capable of valuing a subset of the full range of services, and over time, as new methods are developed, it is expected that a more robust ecosystem services based evaluation framework will emerge.

As noted in Chapter 2, these changes are not anticipated to take effect until after revisions to the accompanying detailed interagency guidelines are released, and congressional action has so far blocked USACE implementation of the *Principles and Requirements*.

In principle, benefit-cost analysis should include *all* benefits and costs of investments in coastal risk reduction including changes in the value of ecosystem services, the value of reduction in risk of fatalities or injuries, as well as the reduction in losses to property and infrastructure, and the direct costs of investment. However, it is difficult to quantify or monetize all of the impacts of investment in coastal risk reduction. Some benefits and costs may be relatively small and it might not be worth the investment of resources necessary to analyze these. In cases where benefits or costs are potentially large but it proves too difficult to estimate monetary values, impacts should still be quantified to the extent possible and constraints should be put on what is considered an acceptable outcome.

Valuing Reductions in Potential Loss of Life

With notable exceptions, such as Hurricane Katrina, relatively few people are killed by coastal storms in the United States compared with other natural catastrophes. Normally, transportation infrastructure for moving people from the coastline is sufficient given advanced warning of approaching storms. In fact, a major benefit of investments in advanced warning of approaching storms or improved transportation infrastructure is a reduction in the expected number of fatalities and injuries. Despite the gains in this area, coastal storms still pose a potential for causing fatalities and injuries, and as such are an important consequence of coastal catastrophes that should be included in the accounting of benefits and costs of coastal risk reduction.

Economists estimate the value of the reduction in the risk of fatalities using the concept of the value of a statistical life (VSL). VSL represents a typical person's willingness to pay to reduce the risk of premature mortality (Mishan, 1971); "for example, a mortality risk of 1/50,000 might be valued at $100, producing a VSL of $5 million" (OMB, 2010). Estimates of VSL can be generated by analyzing wage premiums needed to attract workers to risky jobs or other decisions involving risk of fatality (Mrozek and Taylor, 2002; Viscusi and Aldy, 2003). Estimates of VSL can also be generated by asking people survey questions on how they would choose between risk and income (Krupnick et al., 2002).

The use of VSL in federal decision making, particularly in regulatory applications, is widespread and there is extensive literature on its use (Viscusi, 2004). For example, EPA has long used VSL in evaluating the benefits of the Clean Air Act in reducing mortality due to reduced exposure to air pollution (EPA, 2011). VSL estimates vary depending on methods and data used to construct the estimates as well as by income levels of the populations at risk (Viscusi and Aldy, 2003). The Office of Management and Budget provides guidance to agencies on theory and application of VSL (OMB, 2003) and summarizes the values used by various agencies, noting that these values vary "from roughly $1 million to $10 million per statistical life" but mostly fall in the range above $5 million (OMB, 2010). EPA uses a VSL of $6.3 million (2000 dollars), while the Food and Drug

Administration uses $7.9 million (2010 dollars) and the Department of Transportation uses $6.0 million (2009 dollars) (OMB, 2010).

Rather than deal indirectly with the benefits of reduced fatalities through standard risk criteria, VSL calculations allow risk reductions to be included in benefit-cost analysis along with other benefits and costs measured in monetary terms.

Benefit-Cost Analysis in the Context of Long-Lived Projects

Long-lived investments require an explicit consideration of the future, which necessitates a decision on how to compare present versus future benefits and costs. Investments in coastal risk reduction often generate long-lasting benefits in the form of reduced risk or increases in ecosystem services, or costs in the form of ongoing operation and maintenance expenses or reductions in ecosystem services. Economists discount future values to make them comparable to present values. The typical rationale for discounting is that resources can be invested and earn a positive rate of return (interest) so that receiving an equal amount of money in the future is actually worth less than receiving that amount today.

While discounting is standard practice in most business and economic applications, there are open questions about how to aggregate benefits and costs over time for long-lived projects with significant social or environmental consequences, such as investments in coastal risk reduction. First among these is the proper discount rate to use. OMB recommends a discount rate of 7 percent, but this rate results in greatly reduced benefits or costs beyond a few decades. Others have argued that 7 percent is too high for long-term investments, with longer time horizons and uncertainty about future rates of return leading to lower discount rates (Weitzman, 2001; Arrow et al., 2013). Other debates revolve around whether societal decisions that affect future generations should be treated in the same fashion and use the same discount rate as private investment decisions (see, e.g., the debate around discounting in the context of climate change policy between Nordhaus [2007] and Stern [2007, 2008]). Investments in coastal risk reduction do not raise unique issues in regard to discounting, but evaluating the net present value of such investments requires making potentially difficult or controversial decisions such as determining the proper rate of discount to use in project evaluation.

The long-term nature of investments in coastal risk reduction also means that investment and management will be ongoing rather than a one-time decision. Such recurrent decision making calls for some form of adaptive management (discussed in detail in Box 5-1) in which current investments are evaluated not only with respect to how they affect expected net benefits but also whether the investment maintains or opens options, and whether the investment allows for greater learning about future conditions or the effectiveness of alternative approaches, thereby improving future decision making. There is value (called option value or quasi-option value in the economics literature) to maintaining flexibility (i.e., preserving the option to be adaptive) in the face of uncertainty about the future (Arrow and Fisher, 1974). There is also a "value of information" from learning about future conditions before committing to irreversible decisions because better information allows decisions that are better matched to likely conditions (Hanemann, 1989).

Distributional Issues

Investments in coastal risk reduction generate benefits, some of which accrue primarily to those who live in the coastal communities (e.g., projects to reduce the probability of flooding houses and other private property). Some portion of the costs of coastal risk reduction investments is typically paid by federal taxpayers, including those who live far from the coast. The distribution of benefits and costs across different groups in society raises issues about the proper sharing of responsibilities and rewards for risk reduction. Addressing who should pay the costs of coastal risk reduction raises fairness or equity concerns that are not easily answered. Requiring coastal communities to foot the entire bill for investments in risk reduction can place unaffordable burdens on these communities, especially for those with lower or middle incomes. In addition, some benefits from coastal risk reduction generate widespread benefits that go well beyond just the residents of the coastal community being protected, such as recreation or tourism benefits for nonresidents. But having taxpayers elsewhere pay for investments that provide primarily local benefits is also potentially unfair, especially if taxes come from lower- or middle-income taxpayers and go to wealthy coastal communities. Although there are some general principles that can be applied, such as trying to align costs with beneficiaries of coastal risk reduction, there is typically no simple right answer to distributional issues, and often it is up to the political process to sort out competing claims about what is fair.

AN INTEGRATED RISK-CONSTRAINED BENEFIT-COST APPROACH

The preceding two sections have laid out two coherent but different approaches to evaluating investments in coastal risk reduction—a risk-standard approach and a benefit-cost approach. Although each approach has considerable appeal and numerous examples of application, each approach also has at least one significant weakness. Because the risk-standard approach does not typically factor in benefits other than risk reduction benefits, such as ecosystem services, or explicitly consider costs, this approach may result in choosing investments that yield considerably lower net societal benefits than would alternative investments decisions. The benefit-cost approach, on the other hand, faces the daunting challenge of trying to measure all environmental and social impacts in monetary terms. If such values cannot be accurately measured in monetary terms, then the resulting benefit-cost analysis will be incomplete and misleading.

Given the limitations with each approach, there is an advantage of not rigidly adhering to either approach in its purest form but instead incorporating some elements from each and adopting a hybrid risk-constrained benefit-cost approach. This hybrid approach retains the emphasis on choosing investments that increase net benefits, as in benefit-cost analysis, but puts constraints on what is considered as an acceptable outcome (see, e.g., Figure 4-1). These constraints may arise from societal views on unacceptable risks to which individuals or groups should not be exposed, considerations of equity, or other concerns. Coastal risk planning currently under way in the Netherlands (Box 4-1) represents an example of a hybrid approach that accounts for benefits and costs of investment but adds constraints based on acceptable fatality risk.

A risk-constrained benefit-cost approach is similar in spirit to conclusions of Arrow et al. (1996), who in principle favor the use of benefit-cost analysis in analyzing environmental, health, and safety regulations but are well aware of the practical difficulties of implementing benefit-cost analysis:

> Most economists would argue that economic efficiency, measured as the difference between benefits and costs, ought to be one of the fundamental criteria for evaluating proposed environmental, health, safety regulation. Because society has limited resources to spend on regulation, benefit-cost analysis can help illuminate the trade-offs involved in making different kinds of social investments. . . . In practice, however, the problem is much more difficult, in large part because of inherent problems in measuring marginal benefits and costs. . . . [N]ot all impacts can be quantified, let alone be given a monetary value. Therefore, care should be taken to assure that quantitative factors do not dominate qualitative factors in decision-making. If an agency wishes to introduce a "margin of safety" into a decision, it should do so explicitly.

This hybrid approach is also similar in spirit to OMB and OSTP guidelines on use of risk analysis in federal agencies (Dudley and Hays, 2007). These guidelines recommended that

BOX 4-1
The Coastal Risk Approach in the Netherlands: Past, Present, and Future

Until 1953, coastal protection in the Netherlands was in the hands of 2,600 local water boards, which grew out of medieval grassroots democratic organizations. In 1953 a large flood with a return period of about 250 years overwhelmed the coastal levees in the southwest of the Netherlands with more than 1,800 casualties and an economic loss of 10 percent of gross domestic product. As a result of this catastrophe, coastal risk began to be treated in a more rational and uniform way.

The First Delta Commission advised in 1960 that flood protection levels should be determined based on the value of the property to be protected and the cost of protection (i.e., benefit-cost analysis) (Deltacommissie, 1962). This protection level was cast in terms of a probability of flood for different regions in Holland, ranging from a probability of 1/10,000 per year in central Holland (protecting most major cities) to 1/1,250 (river floodplains) (Figure 4-1-1).

The Commission did consider the value of "loss of life" and other difficult-to-measure values, which were added as a multiplication factor over the real estate value. The Commission factored in land subsidence due to oxidation and the effect that closure to the estuaries would have on the surrounding coast, but did not consider sea-level rise because this was not a known issue in the 1960s. The proposed protection levels remain the law of the land.

As of 2013, the floodprone areas, which constitute about 60 percent of the total land surface of the country (34,000 km^2), are protected with 95 dike rings with a total length of 3,700 km of dunes and (primary) levees. The protected area includes about 10 million people and 2,000 billion Euros of investments. The 650 km of sea and estuarine coasts is protected by about 15 dike rings, and 27 other "sea defenses," such as a closure dam, smaller dams, storm surge barriers, and sluices (Kind, 2013; Ministerie van Verkeer en Waterstaat, 2007).With such a large portion of the country, property, and population in floodprone areas, the Dutch flood risk reduction strategy has been to prevent flooding outright, hence the high levels of protection. In fact, evacuation plans for the larger urban areas do not currently exist as they are deemed not viable logistically.

Recent and Future Developments

Because of the combination of sea-level rise, soil subsidence, increased river runoff, and economic and population growth, flood risk is expected to increase quickly in the Netherlands—in some areas by 4 to 8 percent annually. By the end of the century, this amounts to a flood risk increase by a factor of 30 to 700. This means that the protection standards based on the situation of 1960 are no longer tenable. In 1995, high discharges of the Rhine and Meuse Rivers led to evacuation of more than 250,000 people because of fear the levees would break, which was a wake-up call to the nation. To address these issues, the Dutch government commissioned the Second Delta Commission in 2008. Apart from changing hazard conditions and socioeconomic development, the commission also took natural and cultural values into account. On that basis, they advised a 10-fold, across-the-board increase in the protection levels for all dike ring areas (Kind, 2013; Deltacommissie, 2008).

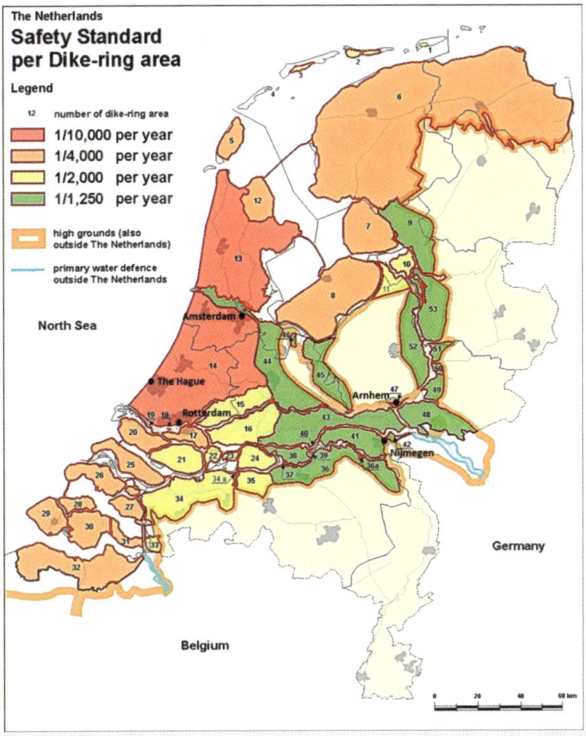

FIGURE.4-1-1 Flood risk criteria for the Netherlands.
SOURCE: Reprinted, with permission, from Van der Most and Wehrung (2005). © 2005 by Natural Hazards.

The Dutch government launched a separate project, Flood Protection for the 21st Century (WV21), which proposed an alternate differentiated approach (Kind, 2013). The WV21 project used benefit-cost analysis and analysis of fatality risk as a basis for proposed new risk reduction standards. Proposed standards are derived on the basis of an optimal dike investment strategy denoting when, where, and how much to invest. The damage cost also includes the cost of human life and other aspects that are difficult to value in monetary terms, with minimal tolerable fatality risks considered separately from total costs and benefits. Additionally, the proposed protection level is no longer cast in terms of a *probability of exceedance* of the water level but in terms of a *probability of flooding*, which takes into account potential levee failures.

This proposed risk framework has not yet been adopted and is the focus of ongoing policy deliberations in the Netherlands, but flood protection improvements are under way.

> Agencies should set priorities for managing risks so that those actions resulting in the greatest net improvement in societal welfare are taken first, accounting for relevant management and social considerations such as different types of health or environmental impacts; individual preferences; the feasibility of reducing or avoiding risks; quality of life; environmental justice; and the magnitude and distribution of both short and long-term benefits and costs.

This hybrid approach is not entirely dissimilar from the current USACE project planning framework, which is constrained by severe environmental impacts (see Chapter 2). However, aside from this constraint, the USACE planning process largely relegates social and environmental factors to levels that do not influence decision making. The USACE approach could be improved through a broader consideration of benefits and costs (as reflected in the Principles and Requirements), including life-safety, environmental, and societal benefits and costs where feasible.

A major challenge with implementing a risk-constrained benefit-cost approach is deciding what categories of coastal risk reduction benefits and costs should be incorporated directly into the benefit-cost calculation, and what categories are best handled by qualitative or nonmonetary quantitative analysis that are incorporated via constraints on what is acceptable versus unacceptable. When constraints are adopted there is also the difficult decision of what outcome levels are viewed as acceptable versus unacceptable.

CONCLUSIONS AND RECOMMENDATIONS

Investments in coastal risk reduction generate significant benefits to society by reducing risk to people and property, but they also involve significant costs. Increases in development and population along the coast, along with sea-level rise, only increase the stakes involved in protecting vulnerable coastal areas. This chapter reviewed two approaches for determining what investments in coastal risk reduction are justified: (1) a risk-standard approach and (2) a benefit-cost approach. Although each approach has considerable appeal, each also has at least one significant weakness. In the case of the risk-standard approach, it is difficult to factor in non-risk-related benefits or costs. In the case of the benefit-cost approach, it is difficult to evaluate all environmental and social impacts in monetary terms. Given the limitations with each approach, there are advantages of not rigidly adhering to either approach in its purest form but instead incorporating some elements from each.

Benefit-cost analysis constrained by acceptable risk and social and environmental dimensions provides a reasonable framework for evaluating

coastal risk management investments. Investments in coastal risk reduction should be informed by net benefits, which include traditional risk reduction benefits (e.g., reduced structural damages, reduced economic disruption) and other benefits (e.g., life-safety, social, environmental benefits), minus the costs of investment in risk reduction and environmental costs. However, because it is difficult to quantify and monetize some benefits and costs, it is important to expand the analysis to include considerations of difficult-to-measure benefits or costs through constraints on what is considered acceptable in social, environmental, and risk reduction dimensions. Such unacceptable levels of risk may include a level of individual risk of fatality, the risk of a large number of deaths from a single event, or adverse impacts on social and environmental conditions that may be difficult to quantify in monetary terms. It is difficult, however, to establish societally acceptable risk standards and requires extensive stakeholder engagement. Setting such a standard requires value judgments, on which not all individuals or groups will necessarily agree.

The recently updated federal guidance for water resources planning—the 2013 *Principles and Requirements for Federal Investments in Water Resources*—provides an effective framework to account for life-safety, social impacts, and environmental costs and benefits in coastal risk reduction decisions. The *Principles and Requirements*, developed by the White House Council on Environmental Quality in response to a 2007 congressional mandate, represents the first step toward federal water resources policy reform. The document, which applies to water resources investment decision making across the federal government—not just within the USACE—recognizes that water resources investment decisions should also consider social and environmental impacts and not give primacy to benefits or costs that are easily measurable in monetary terms. This represents a significant improvement upon current USACE planning, which uses separate accounts for social and environmental impacts, with largely qualitative measures, effectively relegating such considerations to second-class status behind net economic benefits. Progress has been made on measuring improvements in economic terms and on measuring the value of some ecosystem services and social benefits. For other environmental and social factors that are not easily measured in dollar terms, the *Principles and Requirements* recognize that these costs and benefits should also be given adequate weight in decision making. **The Council on Environmental Quality should expedite efforts to complete the detailed accompanying guidelines for implementing the 2013 *Principles and Requirements*, which are required before this framework can to be put into action to improve water resources planning and coastal risk management decision making at the federal agency level.**

Until the updated guidelines to the *Principles and Requirements* are finalized, there are steps the USACE could take to improve consideration of

multiple benefits and costs in the current decision process. Specifically, further attempts in the USACE planning process could be made to more quantitatively consider information in the Environmental Quality and Other Social Effects accounts. For example, work that has been done on how to value ecosystem services could be used to value some environmental quality benefits. Once quantified, these costs and benefits should be rigorously considered and clearly communicated to stakeholders. Such an approach could result in different decision outcomes if the additional social and environmental benefits make certain strategies more acceptable to local sponsors and stakeholders than others. However, trying to quantify or monetize social effects and some environmental effects remains challenging. When only some benefits or costs are monetized there is a tendency to overlook or downplay nonmonetized benefits or costs, and additional attention and/or institutional mechanisms are needed to ensure that these benefits are given adequate weight.

There is no solid basis of evidence to justify a default 1 percent annual-chance (100-year) design level of coastal risk reduction. The 100-year flood criterion used in the National Flood Insurance Program was established for management purposes, not to achieve an optimal balance between risk and benefits. There is also no evidence that reducing risk to a 1 percent annual-chance event is in the best interests of society or that this level is necessarily acceptable to the general public. This level of risk reduction may be appropriate in some settings, unwarranted or excessive in others, and inadequate in highly developed urban areas. Such decisions should, instead, be informed by risk-constrained benefit-cost analyses reflecting site-specific conditions.

5

A Vision for Coastal Risk Reduction

Risks posed by coastal storms are increasing, both to people and to property. As explained in Chapter 1, the growing risk is due both to demographics and to changing natural conditions: population along the coast has expanded and will continue to do so, while sea-level rise and climate change are compounding the threat in the next few decades. Yet, there is no comprehensive national policy on coastal risk that addresses these diverse risk factors. Rather, a complex set of federal, state, and local authorities and agencies addresses these challenges with differing and sometimes conflicting mandates, policies, and approaches. To effectively address the hazards posed by coastal storms, the nation needs a consistent and unified vision for coastal risk reduction. The longer we delay, the more complex the challenge becomes.

The nation's investments in coastal risk reduction are inconsistent and primarily reactive, driven by the latest disaster. As previously described in Chapter 2, the nation readily spends billions of dollars in the wake of disaster, when a significantly smaller investment in mitigation might have averted the calamity in the first place. Congressional authorization for major coastal risk reduction projects occurs when attention is focused on a recent disaster. Once attention fades, the public no longer identifies accumulating coastal risk as a problem worth serious investments, and congressional attention is diverted to other issues. This is inefficient. It falls short of using limited public funds to maximize public safety and to protect property.

A return-on-investment approach to public coastal risk reduction decisions constrained by life-safety and other difficult-to-monetize factors can help rationalize decisions and provide a transparent framework for analyzing alternatives—information necessary to create more resilient coastal communities. This chapter builds upon the information provided in Chapter 4 and provides recommendations for improving the national effort toward a comprehensive vision for reducing coastal risk.

A NATIONAL VISION FOR COASTAL RISK MANAGEMENT

The nation lacks a unifying policy on coastal risks, as it lacks a unifying policy on riverine flood risks. While the concept of a national policy for the coasts is not welcomed by all interests, in part, because it will inevitably have an impact on the status quo, the nation as a whole suffers by its absence. The absence of national policy means that different federal agencies and regional and local jurisdictions plan for and invest in risk reduction in ways that are often inconsistent, leading to inefficient and, too often, inadequate outcomes.

Coastal risk management requires the coordination of efforts that are presently spread across many agencies (see Chapter 2). Today, coastal risk management programs are often uninformed by one another, and their effectiveness is measured against narrow objectives, rather than consistent, overarching national goals. Countervailing policies and federal programs exist that subsidize flood insurance, provide infrastructure investment, and fund emergency response in hazardous areas. Although many federal policies are in place to prevent unwise use of coastal hazard areas, they are biased to maintaining the status quo and often include broad exceptions, grandfathering of previous development, and predisposition to permit post-disaster rebuilding.

Effective coastal risk management necessitates a long-term vision and a comprehensive approach that considers the full array of benefits. Coastal risk management projects have economic and life-safety benefits. They also affect social and ecological systems. Coastal risk management planners, therefore, need to consider this full array of benefits (see Chapter 4) and work collaboratively with related programs, such as housing and development strategies, environmental restoration activities, sustainable economic development programs, and state and local hazard mitigation and adaptation initiatives. In today's coastal management programs, there is limited focus on long-term resilience, planning for future conditions, or comprehensive consideration of nonstructural alternatives for coastal risk reduction. A holistic vision for the coasts would "help ensure continued social, economic, and environmental viability of the nation's precious coastal resources and communities, while minimizing the risks and costs of coastal hazards for present and future generations" (ASFPM Foundation, 2013).

ACHIEVING A NATIONAL VISION

The following steps are key components of developing a national vision for coastal risk management. These build on recent reports of the Association of State Floodplain Managers Foundation (ASFPM Foundation, 2010, 2013), which recommend steps for achieving a national vision for both inland floodplain and coastal risk management.

1. **Establish national objectives for coastal risk reduction.** A key challenge of coastal risk management is the lack of common goals for coastal planning and risk reduction, based on the diversity of entities involved with differing levels of risk, resources, reward, and responsibility (see Chapter 2). Although planning takes place at a local level, the federal government can work collaboratively with state and local governments to craft a vision for sustainable coastal communities and identify objectives and metrics that can serve as risk reduction targets. These objectives and metrics will help localities determine necessary actions and enable assessments of progress toward this vision.
2. **Assess the nation's coastal risks.** To better understand the nation's coastal risk management challenges and appropriately prioritize federal investments, a national assessment of coastal risks is needed. Such an assessment should be based on a standard method for determining future conditions along the coast and methods for quantifying economic, life-safety, social, and environmental costs and benefits associated with risk management scenarios. The assessment should include a national survey and inventory of present and future coastal conditions along the entire seaboard of the United States, including life, property, and infrastructure at risk; of coastal population and development trends; and of coastal environmental resources. Based on this national assessment and accompanying geospatial analysis, the federal government can weigh the benefits of proactive investments in coastal risk reduction against the consequences of no action and identify high priorities for federal funding. Proactive investments could include coastal risk mitigation projects and efforts to improve local land-use planning and decision making. Such an assessment could be part of a broader U.S. flood damage vulnerability assessment mandated in the Water Resources Development Act (WRDA) of 2007. The components of a national assessment of coastal risk are discussed further below.
3. **Incentivize cost-effective risk management strategies.** A sustained effort by the federal government is needed to build state and local capability to prepare and implement more effective mitigation

strategies and policies. This would include technical assistance and promoting the availability of the best scientific data for decision making. To reduce overall cost, the federal government should increase incentives and remove disincentives for improving coastal risk management and land-use planning at the local level.

4. **Build a collaborative approach with clear delineation of responsibilities.** Support for a national vision for coastal risk reduction requires federal leadership and a consistent, collaborative approach. Such an approach would identify and address contradictory agency programs, so that agencies can leverage other related federal and nonfederal efforts and reduce conflicts. The Sandy Rebuilding Task Force of 2013 was an important step in this direction, although the Task Force itself was short-lived. In addition, Executive Order 11988 (1977) ordered federal agencies to minimize actions that result in "adverse impacts associated with the occupancy and modification of floodplains and to avoid direct or indirect support of floodplain development wherever there is a practicable alternative." If implemented comprehensively across agencies, this executive order could provide a baseline for a common and more unified approach.

   To ensure efficient coastal investments, risk reduction and management responsibilities need to be clearly delineated among federal, state, and local agencies, ideally with legislated authorities that clearly lay out these shared and complementary responsibilities. Given the many agencies currently involved in coastal risk reduction, improved federal, state, and local collaboration in support of a national vision will likely necessitate a national-level body for coordinating coastal risk management with participation from all levels. The Federal Interagency Task Force for Floodplain Management, the Mitigation Framework Leadership Group,[1] and the new Council for Climate Preparedness and Resilience (see Chapter 2) are possible models for this body, but any future effort should be cognizant of the deficiencies of prior approaches. Such a collaborative body could also involve professional associations, nongovernmental organizations, and the private sector to foster effective coastal risk reduction.

Developing a vision for coastal risk management is a national prerogative; it is not a responsibility of any single agency alone to create or implement this national vision. The U.S. Army Corps of Engineers (USACE), Federal Emergency Management Agency (FEMA), National Oceanic and Atmospheric Administration (NOAA), Department of Housing and Urban

[1] See http://aswm.org/pdf_lib/nffa/mitigation_framework_leadership_group_charter.pdf.

Development (HUD), U.S. Geological Survey (USGS), and other agencies all have a role to play, as do state and local governments and Congress.

National Coastal Risk Assessment

A national coastal risk assessment is central to a national vision for coastal risk management. Federal agencies should leverage ongoing activities in comprehensive risk assessments to develop a national coastal risk assessment. To maximize limited national resources, a proactive and comprehensive approach is needed to advance the understanding of risk at the community, regional, and national levels. This will require the analysis of risks to a wide range of physical, social, and natural systems.

Under the Disaster Mitigation Act of 2000, states and local jurisdictions are to develop comprehensive risk assessments used in preparing state and local hazard mitigation plans. These are required for eligibility for federal disaster recovery aid. An annual risk-reporting process could be started by compiling these assessments into a risk report by state and for the nation as a whole. States could do the same by local jurisdiction. Such reporting and national attention could improve the quality of risk assessments in state and local plans. This inventory should be based on a "systems approach" that includes consideration of regional ecosystems, watershed and shoreline processes, socioeconomic factors, and multicommunity vulnerability in addition to factors such as life safety, property, and infrastructure at risk.

Risks should be assessed for individual hazards and multiple hazards. Cumulative measures of probable projected losses should be conducted nationwide. The geographic patterns illustrate where the risks and the benefits of targeted interventions are the greatest. The USACE has taken steps toward this end through the North Atlantic Division Comprehensive Study, due in early 2015. In addition, the recently established USACE Risk Management Center will expand capacities of large-scale risk assessment by unifying methodologies across the agency and bringing state-of-the-art methods into play. The state of North Carolina is currently developing an Integrated Hazard Risk Management program, using geospatial analysis tools and models to compile data on "the area, variability, degree, and possibility of impact" of 15 natural hazards including coastal flooding and hurricane winds.[2]

There is a need to develop replicable and robust baseline metrics that are easily understood and applicable to risk management and planning processes. Once established, a set of baseline risk indicators would provide a useful way to monitor and examine change in risks due to a range of factors (e.g., coastal risk reduction projects, state and local land-use policies, and

[2]See irisk.nc.gov/irisk/About.aspx.

changing hazard exposures induced by sea-level rise). Comparisons could also be made across states, regions, and communities to gauge progress in their efforts to reduce risk. The UN Office for Disaster Risk Reduction (UNISDR, 2014) has proposed a *Resilience Scorecard* for making cities more resilient to disasters. A variant of this approach could provide a foundation for the development of national coastal risk metrics.

A critical objective of the coastal inventory and risk assessment is communicating risks to stakeholders involved in coastal decisions. To improve awareness and understanding of coastal risks, a consortium of federal agencies—including but not necessarily limited to the USACE, FEMA, NOAA, USGS, and HUD—in close collaboration with state and local governments, should prepare a periodic coastal risk report. This report would identify the most vulnerable coastal hot spots to forewarn officials and other stakeholders. Information included in the report should communicate levels of risk to the public, and document how risk is changing over time. The report could include scenarios of major disasters under present conditions and projected estimates of loss given changes in urban growth patterns, risk reduction projects, and hazard exposures induced by climate change. The report should address multiple hazards (e.g., wind, sea level, erosion, surge, wave impact, and inundation), linking FEMA floodmaps with NOAA, USGS, HUD, and state information.

Advancing Tools and Data for Coastal Risk Assessment

As discussed in Chapter 1, assessing coastal risk requires a probabilistic evaluation of the hazard (e.g., coastal flooding and wave attack) and a comprehensive evaluation of the consequences. In most cases, the historical record is not long enough to allow the definition of coastal flood hazards from purely observational data. Delineation of coastal flood hazards relies on models of tides, storm surge, waves, and coastal erosion to convert meteorological statistics into projected flood hazards. The methodology and tools for performing these analyses advanced rapidly following hurricane Katrina with substantial USACE leadership. However, significant uncertainty remains in both the storm statistics and the coastal response, particularly as issues associated with climate change are taken into consideration. The evaluation of consequences is also challenging. FEMA's Hazus program provides a nationally applicable, standardized methodology with models for evaluating potential physical damage, economic loss, and social impacts from earthquakes, flooding, and hurricanes. At the state level, the Florida Public Hurricane Loss Model specializes in the estimation of residential loss in Florida from hurricanes. Full consequence analysis depends on data that are only recently becoming available in limited areas (e.g., first-floor elevation data for coastal buildings). Consequence estimation for ecological systems remains in very early stages. Thus, there is considerable room for improved consequence analysis tools and data to support comprehensive risk assessment.

The value of a national coastal risk assessment will depend on its robustness and accuracy. Although a large number of tools exist to address one or more parts of the risk calculation, many are highly empirical and significant uncertainty remains in the results. For example, existing coastal hazard models are biased toward sandy shorelines, whereas models for other types of coast (e.g., vegetated coasts, rocky coasts, hard-structured coasts) are largely missing. Similarly, models that accurately describe the interconnections between storm surge and back-bay and river flooding are needed. Thus, it is important to continue to develop supporting data sets, methodologies, and models that integrate multiple hazards, where feasible. Given the history of the USACE in coastal issues, they are well positioned to continue to play a significant role in the development of these tools. This work should be pursued in an open manner that partners with and leverages the broader international research community and that enables transparency in the eventual results.

The uncertainty associated with existing data and modeling tools should not be viewed as a reason to delay action. Instead, decision makers should use the best available information and take advantage of learning opportunities and adaptive management (see Box 5-1) to improve future risk management. Federal agencies should also continue to improve data availability and modeling tools that can better inform local coastal risk management decisions and periodically revisit guidance for coastal risk reduction planning, considering information gained.

Federal-State Coordination

Once national goals and objectives for coastal risk reduction are established, increased efforts are needed to build risk management capacity at the state and local levels. Thus, an effective, comprehensive risk management framework will require much more extensive support for collaborative partnerships between the federal government and state, local, and private sectors that are charged with implementation. The federal government already provides technical assistance, data, and tools to assist state and local partners to develop local plans that meet coastal risk reduction goals and should continue to improve upon these efforts. It is important that the best available science and data, including emerging social science related to resilience, be available and communicated effectively.

Multiple federal agencies with differing responsibilities face the challenge of clearly presenting unified information, and the variety of stakeholders who are sources and recipients of risk information, adds complexity.

BOX 5-1
Adaptive Management and Coastal Risk Reduction

Adaptive management provides a structured framework premised on active learning that enables adjustments in risk management as new information is developed (Holling, 1978; Walters, 1986). Adaptive management is both a scientific and participatory process that involves identifying goals and agreeing upon critical uncertainties that need to be addressed to improve future decision making (Table 5-1). Once these uncertainties are identified and prioritized, strategies and actions can be planned and implemented based on conceptual (or mathematical) models describing key drivers of change. The system responses are monitored and evaluated, and this knowledge is then used to adapt and improve future management decisions as needed (Murray and Marmorek, 2004; NRC, 2004b, 2011a; Table 5-1). The order in which these steps are carried out is not always linear, but they provide a staged progression from goal setting to implementation, monitoring, and adjustment of actions, with continuous incorporation of scientific knowledge and dialogue with the public.

Adaptive management is well suited for coastal risk reduction efforts given the incomplete knowledge of how a coastal risk management program will reduce damages from future storms. The framework also lends itself to dealing with complex social-ecological dynamics, which present a challenge to coastal risk management. Although adaptive management is new to coastal risk management, it has been more widely used in other management and planning domains that are challenged by a high degree of uncertainty, such as ecosystem restoration and air pollutant emissions trading (see NRC 2004a; Hess et al., 2012). More recently, researchers have documented incorporation of adaptive management innovations into the next generation of urban plans (Quay, 2010; Godschalk and Anderson, 2012; Berke and Lyles, 2013) and local public health management systems (Hess et al., 2012) to increase local capacity to respond to emerging risks posed by climate change. A common theme with prior applications of the adaptive management process is that not all outcomes may be anticipated, but opportunities exist for learning from desirable and undesirable results.

However, an expanding spectrum of technologies provides means for integrating data sets and visualizing information. For example NOAA's Digital Coast[3] provides a range of useful data and tools and the training needed to use them, including coastal LIDAR elevation data, a sea-level-rise and flooding-impact viewer, and tutorials on climate adaptation.

State and federal agencies can also work together within the authorities of current programs to develop improved models for collaboration. One example of cooperation is the Systems Approach to Geomorphic Engineering (SAGE), a multiagency effort including the USACE, NOAA, FEMA,

[3]See http://www.csc. noaa.gov/digitalcoast/.

TABLE 5-1 Key Elements of Adaptive Management with Application to a Coastal Setting

Key Elements of Adaptive Management (AM)	Example application to coastal setting
1. Stakeholder engagement and interagency collaboration	Development of a stakeholder participation program to ensure engagement throughout the AM process. Stakeholders include representatives of public- and private-sector interests affected by AM decisions, and federal, state, and local agencies with relevant interests or expertise are involved.
2. Establish or refine goals	Planners would engage stakeholders to define the goals of the coastal risk reduction effort.
3. Identify and prioritize decision-critical uncertainties	Key unknowns are identified and prioritized based on the degree to which they could inform future decision making. Uncertainties might include: —Can targeted federal incentives significantly enhance coastal mitigation and reduce overall federal expense? —Can targeted coastal land-use planning increase retreat rather than rebuilding after a major event?
4. Apply conceptual models and develop performance measures	Identification of problems should be grounded in an understanding of major trends and drivers of coastal risk, and assessments of opportunities and threats to desirable future conditions. Specific performance measures are identified to assess system response to coastal risk reduction strategies and track goal achievement.
5. Develop and implement robust and flexible management strategies.	Coastal risk reduction strategies are evaluated and implemented, with an emphasis on robust strategies that are applicable across multiple futures and flexible approaches that can be adapted with new information.
6. Monitor system response and assess data	Information is collected and analyzed to compare the outcomes of the management actions relative to the original goals and assess the causes of unexpected results.
7. Incorporate learning into future decisions.	Based on the findings, coastal risk strategies are adapted to enhance effectiveness.

SOURCES: NRC (2011a), RECOVER (2011).

the Nature Conservancy, the Virginia Institute for Marine Sciences, the University of New Orleans, and the University of Rhode Island. Through SAGE, engineers, physical and environmental scientists, educators, and public policy specialists from the federal government, states, academia, nongovernmental organizations, and the private sector work together to advance the knowledge and application solutions and practices to reduce coastal risk (Dalton, 2013).

Strengthening the Role of Consequence Reduction Strategies

Approaches to reduce the consequences of coastal storms are among the most cost-effective strategies to reduce coastal storm risks in many locations, but localities often find land-use planning strategies difficult to implement. On the one hand, local governments and individual property owners seem to place low priority on local actions to reduce coastal risk. This low priority to act is not necessarily due to a lack of awareness. Risk perception research consistently indicates that key decision makers (e.g., urban planners, building inspectors, public works engineers) are aware of natural hazards, but discount the risk and put a low priority on enabling their local governments to take action (Berke and Lyles, 2013). Local decision makers (unless recently hit by a storm event) are inclined to view natural hazards as a marginal problem that has lower priority compared with more pressing concerns such as jobs, roads, and education (Slovic, 1987). Further, the costs of risk-reducing actions are immediate but the benefits are uncertain and long term. Coastal risk reduction benefits are not visible, like a new school or highway, and may not even accrue during the term of elected officials.

On the other hand, federal agencies place limited attention on motivating state and local governments to implement appropriate land-use strategies in high-risk areas. Federal coastal risk reduction projects continue to be built that enable development and redevelopment in high-hazard areas. Additionally, growing federal post-disaster relief reduces the incentives for communities to take action to reduce future losses (see Chapter 2). Improving coastal risk management requires additional focused efforts to assist the general public and public officials to make choices about development and growth to motivate improved pre-disaster mitigation. The federal government has strong interest in reducing disaster outlays by promoting nonstructural mitigation efforts that reduce risk. Federal and state policy makers can make use of a variety of interventions to influence local behavior, including hazard information, technical assistance (e.g., expert review of local planning activities), incentives, direct investment in relocating severe repetitive-loss properties and growth-inducing public infrastructure, planning requirements, and land-use regulations.

One way to broaden responsibility for risk management is through proactive hazard mitigation planning. Rather than simply reacting to a disaster event, local planning enables at-risk communities to become more resilient—to anticipate, absorb, recover from, successfully adapt to future adverse events, and to build back to be safer, healthier, and more equitable (see NRC, 2012b). Such planning considers a wide range of policy instruments such as zoning, regulations, tools that incentivize sound development (e.g., tax increment financing, density bonuses, transfer of development rights), and public capital investments to replace damaged or aging infrastructure (sewer and water). These are powerful tools available to state and local governments to guide development in the most appropriate locations (Table 5-2). Local hazard mitigation planning also provides additional benefits, including public education, consensus building, and improved coordination (see Box 5-2).

As discussed in Chapter 3, studies consistently indicate that where plans aimed at hazard mitigation have been adopted, they foster robust local hazard mitigation programs and a reduction in property damage in natural disasters (Burby and May, 1997; Nelson and French, 2002; May et al., 1996). But state and local hazard mitigation plans are often poorly crafted. Berke and Godschalk (2009) conducted a meta-analysis of these studies and concluded that few communities have prepared well for hazards. Most plans have a weak factual basis (i.e., risk assessments); unclear goals and objectives; weak policies; and few coordination, implementation, and monitoring mechanisms. The most comprehensive study of state and local mitigation plans produced under the federal Disaster Mitigation Act of 2000 completed to date examined 30 coastal state plans (Berke et al., 2012) and 175 local mitigation plans in six states (Lyles et al., 2013) derived similar conclusions concerning the low to moderate quality.

Findings also indicate that although plans are being successfully implemented, they give limited attention to policies oriented toward land use that would reduce the exposure to coastal hazards. Instead, efforts tend to focus on activities that are viewed as easier to achieve (e.g., emergency services and dunes or hard structures to reduce coastal hazards for existing development) and avoid activities that might generate political opposition or impact economic interests. However, when mitigation efforts are integrated into local comprehensive planning efforts, hazard-related losses significantly decline. Often, hazard mitigation plans are not utilized or incorporated into general community land-use planning and development management activities and thus are isolated from these well-established local institutions. Only 12.4 percent of all possible land-use actions are included in local plans, compared with 51 percent for emergency services, 34 percent for education and awareness, and 34 percent for structural risk reduction measures. Most local hazard mitigation plans overlook opportunities to encourage new development to locate outside of flood hazard areas or to assist home and business owners to relocate to safer sites (Berke et al., 2012; Lyles et al., 2013).

Despite the weaknesses of current mitigation planning, the Disaster Mitigation Act (Box 2-2) offers an existing intergovernmental framework that could serve as a foundation for improving risk management practices. Several steps could be taken to strengthen the plans and their associated land-use strategies. First, stronger incentives for local mitigation planning

TABLE 5-2 Land Use Approaches Useful for Mitigating Natural Hazard Risks

Land-Use Approach	Description
Development Regulations	
Permitted land use	Provision regulating the types of land use (e.g., residential, commercial, industrial, open space) permitted in areas of community; may be tied to zoning code
Density of land use	Provision regulating the density of land use (e.g., units per acre); may be tied to zoning code
Subdivision regulations	Provision controlling the subdivision of parcels into developable units and governing the design of new development (e.g., site stormwater management)
Zoning overlays	Provision related to using zoning overlays that restrict permitted land use or density of land use in hazardous areas; may be special hazard zones or sensitive open-space protection zones
Setbacks or buffer zones	Provision requiring setbacks or buffers around hazardous areas (e.g., riparian buffers and ocean setbacks)
Cluster development	Provision requiring clustering of development away from hazardous areas, such as through conservation subdivisions
Density Transfer Provisions	
Density transfer	Provision for transferring development rights to control density; may be transfer of development rights or purchase of development rights
Financial Incentives and Penalties	
Density bonuses	Density bonuses such as ability to develop with greater density in return for dedication or donation of land in areas subject to hazards
Tax abatement	Tax breaks offered to property owners and developers who use mitigation methods for new development
Special study	Provision requiring impact fees or special study fees on development in hazardous areas; may indicate fees required to cover costs of structural risk reduction measures
Land Use Analysis and Permitting Process	
Land suitability	Hazards are one of the criteria used in analyzing and determining the suitability of land for development

TABLE 5-2 Continued

Site review	Provision requiring addressing hazard mitigation in process of reviewing site proposals for development
Public Infrastructure Locations	
Site public facilities	Provision siting new public facilities and replacing and relocating aging facilities out of hazardous areas to steer development to safer locations and to improve prospects to maintain critical services during and after hazard events
Post-Disaster Reconstruction Decisions	
Development moratorium	Provision imposing a moratorium on development for a set period of time after a hazard event
Post-disaster land use change	Provision related to changing land-use regulations following a hazard event; may include redefining allowable land uses after a hazard event
Post-disaster capital	Provisions for relocating and structurally strengthening damaged infrastructure after a disaster.

SOURCE: Reprinted, with permission, from Lyles et al. (2013). © 2013 by Taylor & Francis.

could motivate local jurisdictions to limit or avoid new development or relocate existing development in known hazard areas. Incentives for planning that support land-use actions could be increased in several ways:

- FEMA could increase incentives under the National Flood Insurance Program's Community Rating System (CRS) for communities that adopt a local mitigation plan that accounts for land-use activities or that increase these activities in existing mitigation plans. As of 2013, only 43 percent of local governments that participate in the CRS produced a plan that received credit for flood insurance rate reduction (FEMA, 2013). Additionally, FEMA could give local governments credit under CRS for integrating land-use activities into local comprehensive land-use plans.
- The federal government could link cost sharing for coastal risk reduction to the application of other nonstructural strategies at a local or state level. Under this strategy, local governments (or states) that have progressive public and private property acquisition and relocation programs could pay a smaller share of the cost for federal coastal risk reduction projects. The share of costs could be further decreased if local government were to impose stronger zoning and subdivision restrictions that limit development densities and apply strict building codes in privately owned open spaces in hazardous areas.

BOX 5-2
Benefits of Local Planning for Risk Reduction

Godschalk et al. (1998) describe many benefits of local planning for hazard mitigation. Specifically, hazard mitigation planning:

1. Provides a systematic approach to gathering facts about hazards, the adequacy of existing hazard mitigation policy tools adopted by the community, and a variety of other tools;
2. Educates the community in the course of generating information necessary for decision making, and particularly those with a stake in the outcomes of plans;
3. Demonstrates the connection between the public interest and governmental policies that is critical for legal defensibility;
4. Fosters debate about the issues, and helps build consensus on a vision of resiliency, goals, and action;
5. Coordinates the actions of various federal, state, and local government agencies that affect vulnerability to foster synergy, and avoid duplication of effort and conflict;
6. Guides day-to-day decisions of public officials in the context of broader vision and goals;
7. Provides a means of implementing policy by serving as a reference for elected and appointed officials to use in reaching decisions about regulations, allocating funds for capital investments, and granting permits for development; and
8. Supports monitoring and evaluation of the performance of risk reduction practices based on measurable indicators to gauge goal achievement.

A second step to strengthen local mitigation planning is to build commitment for land-use strategies for reducing coastal risk. Commitment is the willingness of public officials and their constituencies to work energetically to address issues posed by coastal hazards before—not just after—a disaster occurs. Lack of commitment has been a major obstacle to proactive coastal planning and risk reduction, and previous efforts by the federal government to foster local attention to hazards have produced limited commitment. Instead, local officials are more likely to prepare plans to simply comply with the minimum requirements of the Disaster Mitigation Act to be eligible for federal disaster assistance funds rather than create strong plans that integrate mitigation into general community land-use planning and development management activities. State and federal government should do more to engage the public and build commitment from local government officials through public education tools, training, and incentives. Public involvement in the preparation, revision, and updating of mitigation plans and regulations (and other development management measures) can generate understanding and agreement on problems and ways of solving them. Stakeholder engagement efforts give the public a sense of ownership of mitigation proposals and can also foster the formation of coalitions that can work to ensure that permit decisions for development projects are consistent with local mitigation plans (Brody et al., 2003; Godschalk et al., 2003).

The Role of the USACE in a National Vision

The USACE role in coastal risk management is constrained by authorizations that have traditionally emphasized single-project purposes and by the administration, Congress, and appropriations committees that each seek to maintain traditional privileges to authorize and fund specific projects. Nonetheless, as discussed in Chapters 2 and 4, changes are under way that, if implemented, could expand the USACE hurricane and storm damage reduction mission to be more comprehensive in scope. WRDA 2007 directed that the *Principles and Guidelines* (WRC, 1983), which have guided water resources project formulation in multiple agencies since 1983, be revised to include consideration of risk, public safety, and broad social and environmental benefits and include regional planning and nonstructural measures. The first step toward this revision—the *Principles and Requirements for Federal Investments in Water Resources* (CEQ, 2013)—was released in 2013. However, the detailed associated guidelines that will provide instructions for implementing these changes have not been released and are required prior to adoption of this new guidance. As discussed in Chapter 4, the USACE does not need to wait for these revisions to begin implementing a more holistic framework for coastal risk reduction.

Opportunities for Improving USACE Coastal Risk Reduction Strategies

Within the current USACE planning framework, there are several opportunities for improving the planning and implementation of coastal risk reduction strategies to provide greater benefits and increase local responsibility.

Quantify social and economic benefits. More rigorous accounting of social and environmental benefits and costs and life-safety benefits are feasible within the current USACE planning framework. Such analyses (see Chapter 4) would provide greater transparency about the broad costs and benefits of USACE projects and could be used to raise awareness of the value of increased community resilience, social benefits, and ecosystem services that some project alternatives provide.

Incentivize effective coastal planning. Since the reforms of WRDA 1986, local sponsors typically share in the costs of USACE coastal risk reduction projects[4] (see Box 2-3) and enter into local project cooperation agreements (PCAs) that incorporate these cost-sharing arrangements and other conditions. These conditions, however, do not take into consideration the adequacy of the local sponsor's coastal, land-use, and hazard mitigation planning efforts. In fact, perversely, a coastal risk reduction project for a community that has increased development and exposure to risk will have a higher benefit-to-cost ratio than one in a community that has taken action to protect natural features or limit development so that there are fewer people and structures at risk. Under the current decision framework, a USACE risk reduction project for the risk-taking community would have a higher likelihood of funding, thus incentivizing risky development. However, if federal cost-sharing could be made contingent (through PCAs or some other mechanism) upon meeting specific standards for stand-alone coastal hazard mitigation plans and integration of mitigation into local land-use plans, federal investments and cost sharing in coastal risk reduction projects could serve as positive incentives for local communities to reduce exposure to risk.

Embrace long-term coastal planning. Given the long-term challenge of coastal risk reduction in the context of increasing sea-level rise, the typical 50-year USACE planning horizon appears too short to support sound coastal risk management. USACE planners already consider sea-level rise in all coastal projects (USACE, 2013e), but rates of sea-level rise are expected to increase significantly in the latter half of the 21st century (IPCC, 2013), which could significantly impact the effectiveness of coastal risk reduction projects. Unless long-term sea-level rise is considered in all aspects of project planning, coastal risk reduction projects might be selected that spur near-term development and increase long-term exposure to flooding, ultimately increasing overall coastal risks. A planning horizon of 100 years would allow decision makers to consider the adaptability and long-term costs and benefits (including social and environmental effects) of coastal risk reduction alternatives in the context of various sea-level rise projections.

Identify opportunities for learning through adaptive management. Uncertainties regarding the rates of sea-level rise and future changes in hurricane intensity necessitate ongoing improvements in coastal risk management. Therefore, the USACE should embrace adaptive management within its coastal risk reduction efforts so that future decision making can benefit from ongoing learning (see Box 5-1). Adaptive management requires increased effort to identify key uncertainties and monitor outcomes, and not

[4]One major exception to the cost-sharing responsibilities is when Congress adopts an emergency spending bill after a coastal storm to rebuild or construct new storm risk reduction measures at 100 percent federal expense.

all projects are appropriate for this additional level of investment. However, by analyzing uncertainties that currently limit coastal risk management decisions, adaptive management efforts can be targeted so the investments generate knowledge that improves future decision making. To make the most of advances in knowledge, the USACE should, where feasible, design current coastal risk reduction projects with additional flexibility so that the projects can be adapted in the future if needed.

Future Opportunities for Improving USACE Coastal Risk Reduction

Once the detailed *Guidelines* are completed by Council on Environmental Quality (CEQ) to accompany the 2013 *Principles and Requirements* and are formally adopted as guidance for federal water resources planning, additional opportunities will emerge for applying the benefit-cost framework, constrained by acceptable risk, discussed in Chapter 4. Specifically, the *Principles and Requirements*, once implemented, would make it feasible for investments in coastal risk reduction to be informed by net benefits, including traditional risk reduction benefits along with life-safety, social, and environmental benefits, minus the costs of investment and other environmental or social costs. Difficult-to-measure benefits or costs could still be considered through constraints on what is judged to be acceptable.

CONCLUSIONS

A national vision for coastal risk management is needed if comprehensive coastal risk reduction is to be achieved. Effective coastal risk management for the United States requires a national perspective to achieve the most benefits from federal investments and regional solutions, rather than piecemeal, project-by-project approaches. Coastal risk management requires a long-term vision, recognition of the wide array of potential benefits, and coordination of efforts that are currently spread across many agencies that sometimes operate under conflicting mandates. Developing and implementing a national vision for coastal risk management is not the responsibility of any single agency alone, but will require federal leadership and extensive collaboration among federal, state, and local agencies.

The federal government, working closely with states, should establish national objectives and metrics of coastal risk reduction. Specific metrics for coastal risk management could be used by state and local governments to identify necessary actions and assess progress.

The federal government should work with states to develop a national coastal risk assessment. The geographic patterns of disaster risk represented by human fatalities, economic losses, and social impacts can illustrate where the risks are greatest and in need of targeted risk reduction interventions. This analysis should not be based merely on the recent history of hazards but on a comprehensive assessment of risk, including multiple types of hazards under current and anticipated future conditions. The results of the risk assessment would serve as a powerful communication tool for the public and for local and national decision makers. The national interest in coastal risk reduction may vary from one community to another, but this would not preclude a community from investing in risk reduction efforts. The risk assessment would serve as a basis to assess the economic, life-safety, social, and environmental costs and benefits under various risk management scenarios, although additional model development is needed to fully support such an effort.

Stronger incentives are needed to improve pre-disaster risk management planning and mitigation efforts at the local level. Hazard mitigation and adaptation planning has significant potential to reduce coastal risk, but most state and local mitigation plans are currently poor and give limited attention to land-use strategies. In light of behavioral and cognitive factors associated with low-probability, high-consequence events, additional focused efforts and stronger incentives (or disincentives for inaction) are necessary to improve the quality of these plans and the breadth of nonstructural mitigation strategies considered. For example, the federal government could adjust USACE cost sharing for coastal risk reduction projects according to the extent and quality of hazard mitigation planning and the degree to which mitigation is incorporated into other local planning efforts (e.g., land use, transportation, critical infrastructure). The potential for strategic incentives to improve development decisions or facilitate retreat should be carefully examined in the context of long-term cost savings. Federal and state governments should also work to build commitment to coastal risk reduction among stakeholders and local officials.

The USACE should seize opportunities within its existing authorities to strengthen coastal risk reduction. Although the USACE is limited in its capacity to independently initiate national coastal risk reduction strategies under its current authorities, it can use its existing planning framework to rigorously account for social and environmental costs and benefits, thereby supporting a more holistic view of coastal risk management. Additionally, the USACE should increase incentives for sound coastal planning and continue to develop and improve modeling tools to support state and local planning efforts. The USACE should also look for opportunities to apply adaptive management to enhance learning and improve coastal risk reduction strategies. The USACE should reevaluate its typical 50-year planning horizon and consider longer-term planning in the context of projected increases in sea level to assess the adaptability and long-term costs and benefits (including social and environmental effects) associated with risk reduction alternatives.

References

Ackerman, F., and L. Heinzerling. 2002. Pricing the priceless: Cost-benefit analysis of environmental protection. University of Pennsylvania Law Review 150: 1553-1584.

Aerts, J.C. J. H., W. J. Wouter Bozen, H. de Moel, and M. Bowman. 2014. Cost Estimates for flood resilience and protection strategies in New York City. Annals of the New York Academy of Sciences 1294(1): 1-104. DOI: 10.1111/nyas.12200.

Allen, F. R., A. R. Garlick, M. R. Hayns, and A. R. Taig. 1992. The Management of Risk to Society from Potential Accidents: The Main Report of the UKAEA Working Group on the Risks to Society from Potential Major Accidents: With an Executive Summary. London; New York: Elsevier Applied Science.

Arkema, K. K., G. Guannel, G. Verutes, S. A.Wood, A. Guerry, M. Ruckelshaus, P. Kareiva, M. Lacayo, and J. M. Silver. 2013. Coastal habitats shield people and property from sea-level rise and storms. Nature Climate Change 3: 913-918.

Arrow, K., and A. Fisher. 1974. Environmental preservation, uncertainty, and irreversibility. Quarterly Journal of Economics 88: 312-319.

Arrow, K. J., M. L. Cropper, G. C. Eads, R. W. Hahn, L. B. Lave, R. G. Noll, P. R. Portney, M. Russell, R. Schmalensee, V. K. Smith, and R. N. Stavins. 1996. Is there a role for benefit-cost analysis in environmental, health and safety regulation? Science 272: 221-222.

Arrow, K., M. Cropper, C. Gollier, B. Groom, G. Heal, R. Newell, W. Nordhaus, R. Pindyck, W. Pizer, P. Portney, T. Sterner, R. S. J. Tol, and M. Weitzman. 2013. Determining benefits and costs for future generations. Science 341: 349-350.

ASFPM (Association of State Floodplain Managers). 2007. Levees: The Double-Edged Sword. ASFMP White Paper. Available at http://www.floods.org/PDF/ASFPM_Levee_Policy_Challenges_White_Paper_021907.pdf.

ASFPM. 2013. Flood Insurance Affordability—Update. Available at http://www.floods.org/ace-files/documentlibrary/2012_NFIP_Reform/Flood_Insurance_Affordability_version_10222013.pdf.

ASFPM Foundation. 2004. Reducing Flood Losses: Is the 1% Chance Flood Standard Sufficient. Report of the 2004 Assembly of the Gilbert F. White National Flood Policy Forum. Madison, WI.

ASFPM Foundation. 2010. Managing Flood Risks and Floodplain Resources, Report of the Third Assembly of the Gilbert F. White National Flood Policy Forum. Washington, DC: ASFPM Foundation.

ASFPM Foundation. 2013. Holistic Coasts: Adaptive Management of Changing Hazards, Risks, and Ecosystems. A Summary Report Based on the 4th Assembly of the Gilbert F. White National Flood Policy Forum. Available at http://www.asfpmfoundation.org/pdf_ppt/ASFPM-Foundation_HolisticCoasts_Forum2013_Web_Version.pdf.

Associated Press. 1984. Seawall prevents "damage" by Alicia. The Victoria Advocate. March 7, 1984.

Basco, D. R. 2006. Seawall impacts on adjacent beaches: Separating fact from fiction. Journal of Coastal Research SI39: 741-744.

Beachler, K. E., and D. W. Mann. 1996. Long range positive effects of the Delray beach nourishment program. Pp. 4613-4620 in Coastal Engineering 1996. New York: American Society of Civil Engineers.

Bell, S. S., M. O. Hall, and M. L. Middlebrooks. 2014. The value of long-term assessment of restoration: Support from a seagrass investigation. Restoration Ecology. 22(3): 304-310.

Ben C. Gerwick, Inc. 2012. New Orleans East Land Bridge Study. Prepared for the Southeast Louisiana Flood Protection Authority—East, New Orleans, LA. Available at http://www.slfpae.com/presentations/2011-006-001%20-%20Feasibility%20study%20-%20final.pdf.

Bender, M. A., T. R. Knutson, R. E. Tuleya, J. J. Sirutis, G. A. Vecchi, S. T. Garner, and I. M. Held. 2010. Modeled impact of anthropogenic warming on the frequency of intense Atlantic hurricanes. Science 327: 454-458.

Berke, P., and D. Godschalk. 2009. Searching for the good plan: A meta-analysis of plan quality studies. Journal of Planning Literature 23(3): 227-240.

Berke, P., and W. Lyles. 2013. Public risks and the challenges to climate adaptation: A proposed framework for planning in the age of uncertainty. Cityscape: Journal of Policy Development and Research 15(1): 189-216.

Berke, P., D. Godschalk, and E. J. Kaiser. 2006. *Urban Land Use Planning, Fifth Edition.* Urbana: University of Illinois Press.

Berke, P., G. Smith, and W. Lyles. 2012. Planning for resiliency: An evaluation of state hazard mitigation plans under the Disaster Mitigation Act. Natural Hazards Review 13(2): 139-149.

Berke, P., W. Lyles, and G. Smith. 2014. Impacts of federal and state hazard mitigation policies on local land use policy. Journal of Planning Education and Research. doi: 10.1177/0739456X13517004.

Bertness, M. D., and R. Callaway. 1994. Positive interactions in communities. Trends in Ecology and Evolution 9: 191-193.

Bilkovic, D., and M. Roggero. 2008. Effects of coastal development on nearshore estuarine nekton communities. Marine Ecology Progress Series 358: 27-39.

Bilodeau, A. L., and R. P. Bourgeois. 2004. Impact of beach restoration on the deep-burrowing ghost shrimp *Callichirus islagrande*. Journal of Coastal Research 20: 931-936.

Bishop, M. J., C. H. Peterson, H. C. Summerson, H. S. Lenihan, and J. H. Grabowski. 2006. Deposition and long-shore transport of dredge spoils to nourish beaches: Impacts on benthic infauna of an ebb-tidal delta. Journal of Coastal Research 22: 530-546.

Blake, E. S., C. W. Landsea, and E. J. Gibney. 2011. The Deadliest, Costliest and Most Intense United States Tropical Cyclones from 1851 to 2010 (and Other Frequently Requested Hurricane Facts). NOAA Technical Memorandum NWS NHC-6. National Oceanic and Atmospheric Administration. Available at http://www.nhc.noaa.gov/pdf/nws-nhc-6.pdf.

Bocamazo, L. M., W. G. Grosskopf, and F. S. Buonuiato. 2011. Beach nourishment, shoreline change, and dune growth at Westhampton Beach, New York, 1996-2009. Journal of Coastal Research SI59: 181-191.

Bouma, T. J., M. B. De Vries, G. Peralta, I.C. Tanczos, J. van de Koppel, and P. M. J. Herman. 2005. Trade-offs related to ecosystem engineering: A case study on stiffness of emerging macrophytes. Ecology 86: 2187-2199

Bouma, T. J., J. van Belzen, T. Balke, Z. Zhu, L. Airoldi, A. J. Blight, A. J. Davies, C. Galvan, S. J. Hawkins, S. P. G. Hoggart, J. L. Lara, I. J. Losada, M. Maza, B. Ondiviela, M. W. Skov, E. M. Strain, R. C. Thompson, S. Yang, B. Zanuttigh, L. Zhang, and P. M. J. Herman. 2014. Identifying knowledge gaps hampering application of intertidal habitats in coastal protection: Opportunities and steps to take. Coastal Engineering 87: 147-157.

Brody, S., and W. Highfield. 2013. Open space protection and flood mitigation: A national study. Land Use Policy 32: 89-95.

Brody, S., D. Godschalk, and R. Burby. 2003. Mandating citizen participation in plan making: Six strategic planning choices. Journal of the American Planning Association 69(3): 245-264.

Broekx, S., S. Smets, I. Liekens, D. Bulckaen, and L. De Nocker. 2011. Designing a long-term flood risk management plan for the Scheldt estuary using a risk-based approach. Natural Hazards 57: 245-266.

Brooks, R. A., C. N. Purdy, S. S. Bell, and K. J. Sulak. 2006. The benthic community of the eastern US continental shelf: A literature synopsis of benthic faunal resources. Continental Shelf Research 26: 804-818.

Brouwer, R., and D. Pearce (eds.). 2005. Cost-Benefit Analysis and Water Resource Management. Northampton, MA: Edward Elgar.

Browder, A. E., and R. G. Dean. 2000. Monitoring and comparison to predictive models of the Perdido Key beach nourishment project, Florida, USA. Coastal Engineering 30(2-4): 173-191.

Brown, J. T. 2012. FEMA's Community Disaster Loan Program: History, Analysis, and Issues for Congress. Congressional Research Service R42527. Available at https://www.fas.org/sgp/crs/homesec/R42527.pdf.

Browne, M. A., and M. G. Chapman. 2011. Ecologically informed engineering reduces loss of intertidal biodiversity on artificial shores. Environmental Science & Technology 45: 8204-8207.

BSI (British Standards Institution). 2002. Risk Management: Vocabulary—Guidelines for Use in Standards. Published Document ISO/IEC Guide 73:2002.

Building Seismic Safety Council. 1995. NEHRP Recommended Provisions for Seismic Regulations for New Buildings and Other Structures, Part 2—Commentary (FEMA 303), 1994 ed. FEMA 222A Report. Washington, DC: Federal Emergency Management Agency.

Bulleri, F., and M. G. Chapman. 2010. The introduction of coastal infrastructure as a driver of change in marine environments. Journal of Applied Ecology 47: 26-35.

Burby, R. 2006. Hurricane Katrina and the paradoxes of government disaster policy: Bringing about wise governmental decisions for hazardous areas. Annals of the American Academy of Political and Social Science 604: 171-191.

Burby, R. J., and P. J. May. 1997. Making Governments Plan: State Experiments in Managing Land Use. Baltimore, MD: Johns Hopkins University Press.

Burby, R., S. Bollens, E. Kaiser, D. Mullan, and J. Sheaffer. 1988. Cities Under Water: A Comparative Evaluation of Ten Cities' Efforts to Manage Floodplain Land Use. Boulder: Institute of Behavioral Science, University of Colorado.

Burkett, V., and M. Davidson (eds.). 2013. Coastal Impacts, Adaptation and Vulnerabilities: A Technical Input to the 2013 National Climate Assessment. Available at http://downloads.usgcrp.gov/NCA/technicalinputreports/Burkett_Davidson_Coasts_Final_.pdf [accessed June 30, 2014].

Burt, J., A. Bartholemew, and P. Sale. 2011. Benthic development on large-scale engineered reefs: A comparison of communities among breakwaters of different age and natural reefs. Ecological Engineering 37: 191-198.

Callaway, R. M. 1995. Positive interactions among plants. Botanical Review 61: 306-349.

Camfield, F. E. 1993. Different views of beachfill performance. Shore and Beach 61(4): 4-8.

Carbognin, L., P. Teatini, A. Tomasin, and L. Tosi. 2010. Global change and relative sea level rise at Venice: What impact in term of flooding. Climate Dynamics 35: 1055-1063.

Carter, N. T. 2012. Federal Involvement in Flood Response and Flood Infrastructure Repair: Storm Sandy Recovery. Congressional Research Service R-42803. Available at http://fas.org/sgp/crs/homesec/R42803.pdf.

Carter, N. T., and C. V. Stern. 2013. Army Corps of Engineers: Water Resource Authorizations, Appropriations, and Activities. Congressional Research Service R-41243. Available at http://fas.org/sgp/crs/misc/R41243.pdf.

CBO (Congressional Budget Office). 2007. Potential Cost Savings from the Pre-Disaster Mitigation Program (September). Available at http://hazardmitigation.calema.ca.gov/docs/09-28-disaster.pdf.

CEQ (Council on Environmental Quality). 2013. Principles and Requirements for Federal Investment in Water Resources. Available at http://www.whitehouse.gov/sites/default/files/final_principles_and_requirements_march_2013.pdf [accessed January 6, 2014].

Chapman, M. G., and A. J. Underwood. 2011. Evaluation of ecological engineering of "armored" shorelines to improve their value as habitat. Journal of Experimental Marine Biology and Ecology 400: 302-313.

Chen, Y., B. Liao, M. Li, B. Chen, Y. Chen, C. Zhong, H. Li, and W. Lin. 2012. Wind-attenuation effect of *Sonneratia apetala* and *Kandelia obovata* plantations. Yingyong Shengtai Xuebao 23(4): 959-964.

Christianen, M. J., J. van Belzen, P. M. Herman, M. van Katwijk, L. P. Lamers, P. J. van Leent, and T. J. Bouma. 2013. Low-canopy seagrass beds still provide important coastal protection services. PLoS ONE 8(5): e62413.

Cialone, M. A., and D. K. Stauble. 1998. Historical findings on ebb shoal mining. Journal of Coastal Research 14: 537-563.

Climate Central. 2012. Surging Seas, Sea Level Rise Analysis. Available at http://sealevel.climatecentral.org [accessed June 30, 2014].

CNRA (California Natural Resources Agency). 2009. 2009 California Climate Adaptation Strategy. Available at http://resources.ca.gov/climate_adaptation/docs/Statewide_Adaptation_Strategy.pdf [accessed June 30, 2014].

Coastal Research Center. 2013. Beach-Dune Performance Assessment of New Jersey Beach Profile Network (NJBPN) Sites at Northern Ocean County, New Jersey After Hurricane Sandy, Related to FEMA Disaster DR-NJ 4086. Unpublished report, Division of Natural Sciences and Mathematics, Richard Stockton College of New Jersey.

Comiskey, J. J. 2005. Overview of flood damages prevented by U.S. Army Corps of Engineers flood control reduction programs and activities. Journal of Contemporary Water Research and Education 130(1): 13-19.

Convertino, M., J. F. Donoghue, M. L. Chu-Agor, G. A. Kiker, R. Muñoz-Carpena, R. A. Fischer, and I. Linkov. 2011. Anthropogenic renourishment feedback on shorebirds: A multispecies Bayesian perspective. Ecological Engineering 37: 1184-1194.

Corotis, R. B. 2011. Conceptual and analytical difference between resiliency and reliability for seismic hazards. Pp. 2010-2020 in Structures Congress 2011, edited by D. Ames, T. L. Droessler, and M. Hoit. Reston, VA: American Society of Civil Engineers. CD-ROM.

Costanza, R., O. Pérez-Maqueo, M. L. Martinez, P. Sutton, S. J. Anderson, and K. Mulder. 2008. The value of coastal wetlands for hurricane protection. AMBIO: A Journal of the Human Environment 37(4): 241-248.

CPRA LA, 2012. Louisiana's Comprehensive Master Plan for a Sustainable Coast. Coastal Protections and Restoration Authority Members.

Crain, A. D., A. B. Bolten, and K. A. Bjorndal. 1995. Effects of beach nourishment on sea turtles: Review and research initiatives. Restoration Ecology 3: 95-104.

Cutter, G. R., R. J. Diaz, J. A. Musick, J. Olney, Sr., D. M. Bilkovic, J. P.-Y. Maa, S.-C. Kim, C. S. Hardaway, Jr., D. A. Milligan, R. Brindley, and C. H. Hobbs III. 2000. Environmental Survey of Potential Sand Resource Sites Offshore Delaware and Maryland. U.S. Department of the Interior, Minerals Management Service. OCS Study 2000-055. Available at http://www.boem.gov/Non-Energy-Minerals/2000-055.aspx.

Dalton, J. 2013. Sandy Shows We Must Use Holistic Approach. Environmental Forum. Washington, DC: Environmental Law Institute.

Davis, A. B. 1961. The Galveston Sea Wall. Shore and Beach 29(2): 6-13.

Day, J. W., N. P. Psuty, and B. C. Perez. 2000. The role of pulsing events in the functioning of coastal barriers and wetlands: Implications for human impact, management and the response to sea level rise. In Concepts and Controversies in Salt Marsh Ecology, M. P. Weinstein and D. Dreeger, eds. Dordrecht, The Netherlands: Kluwer Academic.

Day, J. W., D. F. Boesch, E. J. Clairain, G. P. Kemp, S. B. Laska, W. J. Mitsch, K. Orth, H. Mashriqui, D. J. Reed, L. Shabman, C. A. Simenstad, B. J. Streever, R. R. Twilley, C. C. Watson, J. T. Wells, and D. F. Whingham. 2007. Restoration of the Mississippi River delta: Lessons from hurricanes Katrina and Rita. Science 315: 1679-1684.

De Lillis, M., L. Costanzo, P. M. Bianco, and A. Tinelli. 2004. Sustainability of sand dune restoration along the coast of the Tyrrhenian Sea. Journal of Coastal Conservation 10: 93-100.

Dean, R. G. 1986. Coastal armoring: Effects, principles and mitigation. Pp 1843-1857 in Proceedings of the 20th International Conference on Coastal Engineering, edited by B. L. Edge. New York: American Society of Civil Engineers.

Dean, R. G. 2002. Beach Nourishment: Theory and Practice. River Edge, NJ: World Scientific

Dean, R. G., and C. J. Bender. 2006. Static wave setup with emphasis on damping effects by vegetation and bottom friction. Coastal Engineering 53(2): 149-156.

Dean, R. G., and R. A. Dalrymple. 2002. Coastal Processes with Engineering Applications. New York: Cambridge University Press.

De Lillis M, Costanzo L, Bianco PM, Tinelli A. 2005. Sustainability of sand dune restoration along the coast of the Tyrrhenian Sea. Journal of Coastal Conservation 10: 93-100.

Deltacommissie. 1962. Rapport Deltacommissie [Report of the Delta Committee]. Eindrapport en Interimadviezen [Final Report and Interim Advice] (in Dutch). The Hague: State Printing and Publishing Office.

Deltacommissie. 2008. Working Together with Water: A Living Land Builds for Its Future. Findings of the Deltacommissie 2008. English synopsis. Available at http://www.deltacommissie.com/doc/deltareport_full.pdf [accessed February 24, 2014].

Dennison, W. C. 1987. Effects of light on seagrass photosynthesis, growth and depth distribution. Aquatic Botany 27: 15-26.

DHS (U.S. Department of Homeland Security). 2007. Infrastructure Resiliency Guide: Reduce Your Vulnerabilities and Make Your Infrastructure Stronger, Version 1.0. Technical Report. Available at https://www.hsdl.org/?view&did=4459.

DHS. 2009a. Congressional Budget Justification: FY 2010. Washington, DC. Available at http://www.dhs.gov/xlibrary/assets/dhs_congressional_budget_justification_fy2010.pdf.

DHS. 2009b. Budget-in-Brief Fiscal Year 2010. Washington, DC. Available at http://www.dhs.gov/xlibrary/assets/budget_bib_fy2010.pdf.

DHS. 2010a. Congressional Budget Justification: FY 2011. Washington, DC. Available at http://www.dhs.gov/xlibrary/assets/dhs_congressional_budget_justification_fy2011.pdf.

DHS. 2010b. FY 2011 Budget in Brief. Washington, DC. Available at http://www.dhs.gov/xlibrary/assets/budget_bib_fy2011.pdf].

DHS. 2011a. Congressional Budget Justification: FY 2012. Washington, DC. Available at http://www.dhs.gov/xlibrary/assets/dhs-congressional-budget-justification-fy2012.pdf.

DHS. 2011b. FY 2012 Budget in Brief. Washington, DC. Available at http://www.dhs.gov/xlibrary/assets/budget-bib-fy2012.pdf.

DHS. 2012a. Survey of Hazard Mitigation Planning. Washington, DC: Office of Inspector General. Available at http://www.oig.dhs.gov/assets/Mgmt/2012/OIG_12-109_Aug12.pdf.

DHS. 2012b. U.S. Department of Homeland Security Annual Performance Report: Fiscal Years 2011–2013. Washington, DC. Available at http://www.dhs.gov/xlibrary/assets/mgmt/cfo_apr_fy2011.pdf.

DHS. 2012c. FY 2013 Budget in Brief. Washington, DC. Available at http://www.dhs.gov/xlibrary/assets/mgmt/dhs-budget-in-brief-fy2013.pdf.

DHS. 2013a. Federal Emergency Management Agency Disaster Relief Fund. FY 2013, Congressional Justification. Available at http://www.fema.gov/pdf/about/budget/11f_fema_disaster_relief_fund_dhs_fy13_cj.pdf.

DHS. 2013b. National Mitigation Framework. Available at http://www.fema.gov/national-mitigation-framework.

DHS. 2013c. National Response Framework, 2nd Ed. Available at http://www.fema.gov/national-response-framework.

DHS. 2013d. U.S. Department of Homeland Security Annual Performance Report: Fiscal Years 2012 – 2014. Washington, DC. Available at http://www.dhs.gov/sites/default/files/images/DHS%20FY%202012-FY%202014%20Annual%20Performance%20Report.pdf.

DHS. 2013e. Budget-in-Brief Fiscal Year 2014. Washington, DC. Available at http://www.dhs.gov/sites/default/files/publications/MGMT/FY%202014%20BIB%20-%20FINAL%20-508%20Formatted%20%284%29.pdf.

DHS. 2014a. U.S. Department of Homeland Security Annual Performance Report: Fiscal Years 2013–2015. Washington, DC. Available at https://www.dhs.gov/sites/default/files/publications/MGMT/DHS-FY-2013-FY-2015-APR.pdf.

DHS. 2014b. Budget-in-Brief Fiscal Year 2015. Washington, DC. Available at http://www.dhs.gov/sites/default/files/publications/FY15BIB.pdf.

Diamond, P., and J. Hausman. 1994. Contingent valuation: Is some number better than no number? Journal of Economic Perspectives 8: 45-64.

Dijkman, J (ed.). 2007. A Dutch Perspective on Coastal Louisiana Flood Risk Reduction and Landscape Stabilization: Second Interim Report to the U.S. Army. London: International Research Office of the U.S. Army.

Doody, J. P. 2001. Coastal conservation and management: An ecological perspective. Dordrecht, The Netherlands: Kluwer Academic.

Douglass, S. L., and J. Krolak. 2008. Highways in the Coastal Environment. Publication FHWA-NHI-07-096. U.S. Department of Transportation, Federal Highway Adminsitration. Available at http://www.fhwa.dot.gov/engineering/hydraulics/pubs/07096/07096.pdf.

Douglass, S. L., and B. H. Pickel. 1999. The tide doesn't go out anymore—the effect of bulkheads on urban shorelines. Shore and Beach 67: 19-25.

Duarte, C. M., I. J. Losada, I. E. Hendriks, I. Mazarrasa, and N. Marbà, 2013. The role of coastal plant communities for climate change mitigation and adaptation. Nature Climate Change 3: 961-968.

Dudley, S. E., and S. L. Hays. 2007. Updated Principles for Risk Analysis. Memorandum for the Heads of Departments and Executive Agencies. M-07-24. Office of Management and Budget, and Office of Technology Policy, Washington, DC. Available at http://www.whitehouse.gov/sites/default/files/omb/memoranda/fy2007/m07-24.pdf

Dugan, J. E., and D. M. Hubbard. 2006. Ecological responses to coastal armoring on exposed sandy beaches. Shore and Beach 74(1): 10-16.

Dugan, J. E., D. M. Hubbard, I. F. Rodil, D. L. Revell, and S. Schroeter. 2008. Ecological effects of coastal armoring on sandy beaches. Marine Ecology 29: 160-170.

Eelkema, M., Z. B. Wang, A. Hibma, and M. J. Stive. 2013. Morphological effects of the Eastern Scheldt storm surge barrier on the ebb-tidal delta. Coastal Engineering Journal 55(3): 1350010, doi:10.1142/S0578563413500101.

Emanuel, K. A. 2013. Downscaling CMIP5 climate models shows increased tropical cyclone activity over the 21st century. Proceedings of the National Academy of Sciences of the United States of America, doi: 10.1073/pnas.1301293110.

Environment Agency. 2007. Flood Risk Management Estimating Guide. Unit Cost Database 2007. Bristol, UK: Environment Agency.

EPA (U.S. Environmental Protection Agency). 1988. National Emission Standards for Hazardous Air Pollutants; Benzene Emissions From Maleic Anhydride Plants, Ethylbenzene/Styrene Plants, Benzene Storage Vessels, Benzene Equipment Leaks, and Coke By-Product Recovery Plants, 53 Fed. Reg. 28496.

EPA, Office of Air and Radiation. 2011. The Benefits and Costs of the Clean Air Act from 1990 to 2020. Washington, DC: USEPA.

Ericksen, N. J., P. R. Berke, J. L. Crawford, and J. E. Dixon. 2004. Plan-making for Sustainability: The New Zealand Experience. Aldershot: Ashgate.

Executive Office of the President. 1952. Reports and budget estimates relating to Federal programs and projects for conservation, development, or use of water and related land resources. Bureau of the Budget Circular A-47. December 31, 1952.

Executive Office of the President. 1977. Presidential Executive Order 11988 (Floodplain Management). Washington, DC.

Executive Office of the President. 2013. The President's Climate Action Plan. June 2013. Available at http://www.whitehouse.gov/sites/default/files/image/president27sclimateactionplan.pdf.

Ezer, T., L. P. Atkinson, W. B. Corlett, and J. L. Blanco. 2013. Gulf Stream's induced sea level rise and variability along the U.S. mid-Atlantic coast. Journal of Geophysical Research 118(2): 685-697.

Feagin, R. A. 2005. Artificial dunes created to protect property on Galveston Island, Texas: The lessons learned. Ecological Restoration 23: 89-94.

FEMA (Federal Emergency Management Agency). 2000. Above the Flood: Elevating Your Floodprone Home. Report No. 347. Washington, DC: FEMA. Available at http://www.fema.gov/media-library-data/20130726-1443-20490-7815/fema347_complete.pdf.

FEMA. 2012. Community Rating System Fact Sheet. Washington, DC: FEMA. Available at http://www.fema.gov/media-library/assets/documents/9998.

FEMA. 2013. National Flood Insurance Program Community Rating System Coordinator's Manual. FIA-15/2013. Washington, DC: FEMA. Available at http://www.fema.gov/media-library-data/20130726-1557-20490-9922/crs_manual_508_ok_5_10_13_bookmarked.pdf.

Fenster, M. S., C. B. Knisley, and C. T. Reed. 2006. Habitat preference and the effects of beach nourishment on the federally threatened northeastern beach tiger beetle, *Cicindela dorsalis*: Western shore, Chesapeake Bay, Virginia. Journal of Coastal Research 22: 1111-1144.

Ferrario, F., M. W. Beck, C. D. Storlazzi, F. Micheli, C. C. Shepard, and L. Airoldi. 2014. The effectiveness of coral reefs for coastal hazard risk reduction and adaptation. Nature Communications 5: 3794, doi:10.1038/ncomms4794.

Fewtrell, L., and J. Bartram (eds.). 2001. Water quality: Guidelines, Standards and Health. London: IWA Publishing.

FIFMTF (Federal Interagency Floodplain Management Task Force). 1986. A Unified National Program for Floodplain Management. Washington, DC: Federal Emergency Management Agency. Available at http://www.fema.gov/media-library-data/20130726-1503-20490-9177/fema100.pdf.

FIFMTF. 1989. Conceptual Framework and Basic Strategies and Tools for Implementing a Unified National Program for Floodplain Management. Washington, DC: Federal Emergency Management Agency.

FIFMTF. 1994. A Unified National Program for Floodplain Management. Washington, DC: Federal Emergency Management Agency. Available at http://www.fema.gov/media-library-data/20130726-1733-25045-0814/unp_floodplain_mgmt_1994.pdf.

FIFMTF. 2012. Guidance on "Unwise Use of Floodplains." Washington, DC: Federal Emergency Management Agency. Available at http://www.fema.gov/media-library-data/20130726-1841-25045-7320/fifm_tf_signedunwiseuseguidance_3_26_12.pdf.

FIFMTF. 2013. FIFM Task Force Concensus Recommendations and Actions from a Federal Floodplain Management Policy Analysis. Washington, DC: Federal Emergency Management Agency. Available at http://www.fema.gov/media-library-data/20130726-1905-25045-0495/fifm_tf_consensus_recommendations_signed.pdf.

Finkl, C. W. 1996. What might happen to America's shorelines if artificial beach replenishment is curtailed: A prognosis for southeastern Florida and other sandy regions along regressive coasts. Journal of Coastal Research 12: iii-ix.

FitzGerald, D. M., M. S. Fenster, B. A. Argow, and I. V. Buynevich. 2008. Coastal impacts due to sea-level rise. Annual Review of Earth and Planetary Sciences 36: 601-647.

Foderaro, L. 2013. Breach through Fire Island also divides opinions. New York Times, April 2, A13.

Fonseca, M. S., J. S. Fisher, J. C. Zieman, and G. W. Thayer. 1982. Influence of the seagrass, *Zostera marina* L., on current flow. Estuarine, Coastal and Shelf Science 15(4): 351-364.

Freeman, A. M. III. 2003. The Measurement of Environmental and Resource Values: Theory and Methods. Washington, DC: Resources for the Future.

Freestone, A. L., and K. F. Nordstrom. 2001. Early evolution of restored dune plant microhabitats on a nourished beach at Ocean City, New Jersey. Journal of Coastal Conservation 7: 105-116.

FWS (U.S. Fish and Wildlife Service). 2002. Draft Fish and Wildlife Coordination Act Report: Bogue Banks Shore Protection Project, Carteret County, NC. Raleigh, NC: Ecological Services Field Office, FWS.

Gall, M., K. A. Borden, C. T. Emrich, and S. L. Cutter. 2011. The unsustainable trend of natural hazard losses in the United States. Sustainability (3): 2157-2181.

Galloway, G. E., G. B. Baecher, D. Plasencia, K. G. Coulton, J. Louthain, M. Bagha, and A. R. Levy. 2006. Assessing the Adequacy of the National Flood Insurance Program's 1 Percent Flood Standard. College Park, MD: Water Policy Collaborative, University of Maryland.

Gambi, M. C., A. Nowell, and P. A. Jumars. 1990. Flume observations on flow dynamics in *Zostera marina* (eelgrass) beds. Marine Ecology Progress Series 61(1-2): 159-169.

GAO (U.S. General Accounting Office). 1976. Cost, Schedule, and Performance Problems of the Lake Pontchartrain and Vicinity, Louisiana, Hurricane Protection Project. PSAD-76-161. Washington, DC: GAO.

GAO (U.S. Government Accountability Office). 2005. Hurricane Protection: Statutory and Regulatory Framework for Levee Maintenance and Emergency Response for the Lake Pontchartrain Project. GAO-06-322T. Washington, DC: U.S. Government Printing Office.

GAO. 2007. Coastal Barrier Resources System: Status of Development That Has Occurred and Financial Assistance Provided by Federal Agencies. GAO-07-356. Washington, DC: GAO. Available at http://www.gao.gov/assets/260/257815.pdf.

GAO. 2013. Hurricane Sandy Relief: Improved Guidance on Designing Internal Control Plans Could Enhance Oversight of Disaster Funding. GAO-14-58. Washington, DC: GAO. Available at http://www.recovery.gov/Sandy/Documents/GAO%20Improved%20Guidance%20Sandy.pdf.

Gedan, K., and B. R. Silliman. 2009. Patterns of salt marsh loss within coastal regions of North America. Pp. 253-266 in Human Impacts on Salt Marshes: A Global Perspective, edited by B. R. Silliman, T. Grosholz, and M. D. Bertness. Berkeley: University of California Press.

Gedan, K. B., M. L. Kirwan, E. Wolanski, E. B. Barbier, and B. R. Silliman. 2011. The present and future role of coastal wetland vegetation in protecting shorelines: Answering recent challenges to the paradigm. Climatic Change 106: 7-29.

GFDL (Geophysical Fluid Dynamics Laboratory). 2013. Global Warming and Hurricanes: An Overview of Current Research Results. Available at http://www.gfdl.noaa.gov/global-warming-and-hurricanes.

Godschalk, D., and W. Anderson. 2012. Sustaining Places: The Role of the Comprehensive Plan. Planning Advisory Service Report No. 567. Chicago, IL: American Planning Association.

Godschalk, D., E. Kaiser, and P. Berke. 1998. Land Use Planning and Natural Hazard Mitigation. Pp. 85-118 in Cooperating with Nature: Confronting Natural Hazards with Land Use Planning for Sustainable Communities, edited by R. Burby. Washington, DC: Joseph Henry Press.

Godschalk, D., S. Brody, and R. Burby. 2003. Public Participation in natural hazard mitigation policy: Challenges in comprehensive planning. Journal of Environmental Planning and Management 46(5): 733-754.

Godschalk, D., A. Rose, E. Mittler, K. Porter, and C. T. West. 2009. Estimating the value of foresight: Aggregate analysis of natural hazard mitigation benefits and costs. Journal of Environmental Planning and Management 52(6): 739-756.

Gorzelany, J. F., and W. G. Nelson. 1978. Biological effects of beach nourishment on a sub-tropical beach. Marine Environmental Research 21: 75-94.

Gouldby, B., and P. Samuels. 2005. Integrated Flood Risk Analysis and Management Methodologies: Language of Risk Project Definitions. Report No. T32-04-0. FLOODsite Consortium. Available at http://www.floodsite.net/html/partner_area/project_docs/floodsite_language_of_risk_v4_0_p1.pdf.

Grabowski, J. H, and C. H. Peterson. 2007. Restoring oyster reefs to recover ecosystem services. Pages 281-298 in K. Cuddington, J. E. Byers, W. G. Wilson, and A. Hastings, eds. Ecosystem Engineers: Plants to Protists. Elsevier.

Grafals-Soto, R. 2012. Effects of sand fences on coastal dune vegetation distribution. Geomorphology 145-146: 45-55.

Graham, H. E., and D. E. Nunn. 1959. Meteorological Considerations Pertinent to Standard Project Hurricane, Atlantic and Gulf Coasts of the United States. National Hurricane Research Project Report No. 33. Washington, DC.: Weather Bureau, U.S. Department of Commerce.

Griggs, G.B., J. F. Tait, K. Scott, and N. Plant. 1991. Tbe Interaction of Seawalls and Beaches: Four Years of Field Monitoring. Monterey Bay, California. Proceedings Coastal Sediments, American Society of Civil Engineers.

Griggs, G.B., J. F. Tait, and W. Corona. 1994. The Interaction of Seawalls and Beaches: Seven Years of Monitoring. Monterey Bay, California.Shore and Beach 62(3): 21-28.

Grinsted, A., J. C. Moore, and S. Jevrejeva. 2013. Projected Atlantic hurricane surge threat from rising temperatures. Proceedings of the National Academy of Science of the United States of America, 110: 5369-5373.

Haas, H. L., K. A. Rose, B. Fry, T. J. Minello, and L. P. Rozas. 2004. Brown shrimp on the edge: Linking habitat to survival using an individual-based simulation model. Ecological Applications 14(4): 1232-1247.

Hallegatte, S. 2012. Toward a world of larger disasters? Ideas for risk-management policies. VoxEU, Centre for Economic Policy Research. Available at http://www.voxeu.org/article/toward-world-larger-disasters-ideas-risk-management-policies.

Hallegatte, S., C. Green, R. J. Nicholls, and J. Corfee-Morlot. 2013. Future flood losses in major coastal cities. Nature climate change 3(9): 802-806.

Hammond, R. J. 1966. Convention and limitation in benefit-cost analysis. Natural Resources Journal 6: 195-222.

Hanemann, W. M. 1989. Information and the concept of option value. Journal of Environmental Economics and Management 16: 23-37.

Hanemann, W. M. 1994. Contingent valuation and economics. Journal of Economic Perspectives 8: 19-44.

Hanson, S., R. Nicholls, N. Ranger, S. Hallegatte, J. Corfee-Morlot, C. Herweijer, and J. Chateau. 2011. A global ranking of port cities with high exposure to climate extremes. Climatic Change 104(1): 89-111. DOI: 10.1007/s10584-010-9977-4

Harrison, D., and D. Rubinfeld. 1978. Hedonic housing prices and the demand for clean air. Journal of Environmental Economics and Management 5: 81-102.

Hawkins, S., H. F. Burcharth, B. Zanuttigh, and A. Lamberti. 2007. Environmental Design Guidelines for Low Crested Coastal Structures, 1st Ed. Amsterdam: Elsevier Science.

Hearon, G. E., W. G. McDougal, and P. D. Komar. 1996. Long-term beach response to shore stabilization structures on the Oregon coast. Pp. 2718-2731 in Coastal Engineering 1996. New York: American Society of Civil Engineers.

Hess, J., J. McDowell, and G. Lube. 2012. Integrating climate change adaptation into public health practice: Using adaptive management to increase adaptive capacity and build resilience. Environmental Health Perspectives 120(2): 171-179.

Hettiarachchi, S. L., S. P. Samarawickrama, H. J. S. Fernando, A. Harsha, R. Ratnasooriya, N. A. Kithsiri Nandasena, and S. Bandara. 2013. Investigating the performance of coastal ecosystems for hazard mitigation. Pp. 57-81 in The Role of Ecosystems in Disaster Risk Reduction, edited by F. G. Renaud, K. Sudmeier-Rieux, and M. Estrella. Tokyo: United Nations University Press.

Hobbs, C. H. III. 2002. An investigation of potential consequences of marine mining in shallow water: An example from the mid-Atlantic coast of the United States. Journal of Coastal Research 18: 94-101.

Holling, C. 1978. Adaptive Environmental Assessment and Management. New York: Wiley.

Houston, J. R. 1991. Beachfill performance. Shore and Beach 59(3): 15-24.

Houston, J. R. 1996. Engineering practice for beach-fill designs. Shore and Beach 64(3): 27-35.

Houston, J. R. 2013. The value of Florida beaches. Shore and Beach 81(4): 4-11.

Howe, C. W. 1971. Benefit-Cost Analysis for Water System Planning. Washington, DC: American Geophysical Union.

HSE (Health and Safety Executive). 2001. Reducing Risks, Protecting People: HSE's Decision-making Process. London: HSE Books, Her Majesty's Stationery Office.

Hunter, P. R., and L. Fewtrell. 2001. Acceptable risk. Pp. 207-227 in Water Quality: Guidelines, Standards and Health, edited by L. Fewtrell and J. Bartram. London: IWA Publishing.

Hurricane Sandy Rebuilding Task Force. 2013. Hurricane Sandy Rebuilding Strategy. U.S. Department of Housing and Urban Development. August.

ICCATF (Interagency Climate Change Adaptation Task Force). 2011. Federal Actions for a Climate Resilient Nation: A Progress Report of the Interagency Climate Change Adaptation Task Force. Available at http://www.whitehouse.gov/sites/default/files/microsites/ceq/2011_adaptation_progress_report.pdf.

IOM (Institute of Medicine). 2013. Environmental Decisions in the Face of Uncertainty. Washington, DC: The National Academies Press.

IPCC (Intergovernmental Panel on Climate Change). 2013. Changes to the underlying scientific/technical assessment. In Climate Change 2013: The Physical Science Basis. A Special Report of Working Group I Contribution to the IPCC 5th Assessment Report. Available at https://www.ipcc.ch/report/ar5/wg1/docs/review/P36Doc4_WGI-12_Changes-Underlying-Assessment.pdf.

IPET (Interagency Performance Evaluation Task Force). 2007. Performance Evaluation of the New Orleans and Southeast Louisiana Hurricane Protection System, Vol. 5: The Performance—Levees and Floodwalls. Final Report. Available at https://ipet.wes.army.mil/.

IPET. 2009. Performance Evaluation of the New Orleans and Southeast Louisiana Hurricane Protection System: Final Report. Available at https://ipet.wes.army.mil/ [accessed December 4, 2012].

Irish, J., P. Lynett, R. Weiss, S. Smallegan, and W. Chen. 2013. Buried relic seawall mitigates Hurricane Sandy's impacts. Coastal Engineering 80: 79-82.

Jackson, N. L., K. F. Nordstrom, and D. R. Smith. 2002. Geomorphic-biotic interactions on beach foreshores in estuaries. Journal of Coastal Research SI36: 414-424.

Jackson, N. L., D. R. Smith, and K. F. Nordstrom. 2005. Comparison of sediment grain size characteristics on nourished and un-nourished estuarine beaches and impacts on horseshoe crab habitat, Delaware Bay, New Jersey. Zeitschrift für Geomorphologie Supplementband 141: 31-45.

Jackson, N. L., D. R. Smith, R. Tiyarattanachi, and K. F. Nordstrom. 2007. Evaluation of a small beach nourishment project to enhance habitat suitability for horseshoe crabs. Geomorphology 89: 172-185.

Jackson, N. L., D. R. Smith, and K. F. Nordstrom. 2008. Physical and chemical changes in the foreshore of an estuarine beach: Implications for viability and development of horseshoe crab (*Limulus polyphemus*) eggs. Marine Ecology Progress Series 355: 209-218.

Jackson, N. L., K. F. Nordstrom, S. Saini, and D. R. Smith. 2010. Effects of nourishment on the form and function of an estuarine beach. Ecological Engineering 36: 1709-1718.

Jiabi, X., P. Sayers, S. Dongya, and Z. Hanghui. 2013. Broad-scale reliability analysis of the flood defence infrastructure within the Taihu Basin, China. Journal of Flood Risk Management 6: 42-56.

Jones, H. P., D. G. Hole, and E. S. Zavaleta. 2012. Harnessing nature to help people adapt to climate change. Nature Climate Change 2: 504-509.

Jones, S. C., D. K. Tidwell, and S. B. Darby. 2009. Comprehensive Shoreline Mapping, Baldwin and Mobile Counties, Alabama: Phase 1. Open File Report 0921. Geological Survey of Alabama, Tuscaloosa.

Jones, S. R., and W. R. Magnun. 2001. Beach nourishment and public policy after Hurricane Floyd: Where do we go from here? Ocean and Coastal Management 44: 207-220.

Jonkman, S., M. Kok, and J. Vrijling. 2008. Flood risk assessment in the Netherlands: A case study for Dike Ring South Holland. Risk Analysis 28(5): 1357-1373.

Jonkman, S., R. Jongejan, and B. Maaskan. 2011. The use of individual and societal risk criteria within the Dutch flood safety policy—nationwide estimates of societal risk and policy applications. Risk Analysis 31(2): 282-300.

Jordan, L. K. B., K. W. Banks, and L. E. Fisher. 2010. Elevated sedimentation on coral reefs adjacent to a beach nourishment project. Marine Pollution Bulletin 60: 261-271.

Keeney, R., and H. Raiffa. 1993. Decisions with Multiple Objectives: Preferences and Value Tradeoffs. Cambridge, UK: Cambridge University Press.

Kelley, S. W., J. S. Ramsey, and M. R. Byrnes. 2001. Numerical Modeling Evaluation of the Cumulative Physical Effects of Offshore Sand Dredging for Beach Nourishment. BOEM-2001-098. Washington, DC: Bureau of Ocean Energy Management.

Kelman, S. 1981. Cost-benefit analysis: An ethical critique. AEI Journal on Government and Society Regulation 5(January/February): 33-40.

Kind, J. M. 2013. Economically efficient flood protection standards for the Netherlands. Journal of Flood Risk Management. 7(2): 103-117.

King, R. O. 2013. The National Flood Insurance Program: Status and Remaining Issues for Congress. Congressional Research Service. R42850.

Knutson, T. R., J. J. Sirutis, G. A. Vecchi, S. Garner, M. Zhao, H.-S. Kim, M. Bender, R. E. Tuleya, I. M. Held, and G. Villarini. 2013. Dynamical downscaling projections of late twenty-first century Atlantic hurricane activity: CMIP3 and CMIP5 model-based scenarios. Journal of Climate 26: 6591-6617.

Kobayashi, N., A. W. Raichle, and T. Asano. 1993. Wave attenuation by vegetation. Journal of Waterways Port Coastal Ocean Engineering 119(1): 30-48.

Kraus, N. C. 1988. The effects of seawalls on the beach: An extended literature review. Journal of Coastal Research SI4: 1-29.

Kraus, N. C., and L. Lin. 2009. Hurricane Ike along the upper Texas coast: An introduction. Shore and Beach 77(2): 3-8.

Kraus, N. C., and W. G. McDougal. 1996. The effects of seawalls on the beach: Part I, An updated literature review. Journal of Coastal Research 12(3): 691-701.

Krauss, K. W., T. W. Doyle, T. J. Doyle, C. M. Swarzenski, A. S. From, R. H. Day, and W. H. Conner. 2009. Water level observations in mangrove swamps during two hurricanes in Florida. Wetlands 29(1): 142-149.

Krupnick, A., A. Alberini, M. Cropper, N. Simon, B. O'Brien, R. Goeree, and M. Heintzelman. 2002. Age, health, and the willingness to pay for mortality risk reductions: A contingent valuation survey of Ontario residents. Journal of Risk and Uncertainty 24(2): 161-186.

Kunreuther, H. 1996. Mitigating disaster losses through insurance. Journal of Risk and Uncertainty 12: 171-187.

Kunreuther, H. 2006. Disaster mitigation and insurance: Learning from Katrina. Annals of the American Academy of Political and Social Science 604: 208-227.

Kunreuther, H. 2008. Moral Hazard. Encyclopedia of Quantitative Risk Analysis and Assessment III. DOI: 10.1002/9780470061596.risk0655.

Landsea, C. W. 2007. Counting Atlantic tropical cyclones back to 1900. Eos, Transactions of the American Geophysical Union 88(18): 197-202.

Lavery, S., and B. Donovan. 2005. Flood risk management in the Thames Estuary looking ahead 100 years. Philosophical Transactions of the Royal Society A 363(1831): 1455-1474.

Leatherman, S. 1989. National Assessment of Beach Nourishment Requirements Associated with Accelerated Sea Level Rise. U.S. Environmental Protection Agency Office of Policy, Planning, and Evaluation.

Leewis, L., P. M. van Bodegom, J. Rozema, and G. M. Janssen. 2012. Does beach nourishment have long-term effects on intertidal macroinvertebrate species abundance? Estuarine, Coastal and Shelf Science 113: 172-181.

Lenihan, H.S. 1999. Physical-biological coupling on oyster reefs: How habitat structure influences individual performance. Ecological Monographs 69(3): 251-275.

Lin, N., K. A. Emanuel, M. Oppenheimer, and E. Vanmarcke. 2012. Physically based assessment of hurricane surge threat under climate change. Nature Climate Change 2: 462-467.

Lindeman, K. C., and D. B. Snyder. 1999. Nearshore hardbottom fishes of southeast Florida and effects of habitat burial caused by dredging. Fishery Bulletin 97: 508-525.

Lindsay, B. R. 2014. FEMA's Disaster Relief Fund: Overview and Selected Issues. Congressional Research Service. Available at http://fas.org/sgp/crs/homesec/R43537.pdf.

Lindsay, B. R., and J. Murray. 2011. Disaster Relief Funding and Emergency Supplemental Appropriations. Congressional Research Service R40708.

Linham, M. M., C. H. Green, and R. J. Nicholls. 2010. Costs of Adaptation to the Effects of Climate Change in the World's Large Port Cities. AVOID Report AV/WS2/D1/R14. Available at http://www.metoffice.gov.uk/media/pdf/k/s/AVOID_WS2_D1_14_20100701.pdf.

London, J. B., C. S. Dyckman, J. S. Allen, C. C. St. John, I. L. Wood, and S. R. Jackson. 2009. An Assessment of Shoreline Management Options Along the South Carolina Coast. Clemson University Strom Thurmond Institute of Government and Public Affairs and South Carolina Water Resources Center. Available at http://www.scdhec.gov/environment/ocrm/docs/SCAC/Shoreline_Options_SC.pdf.

Lyles, W., P. Berke, and G. Smith. 2013. Do planners matter? Examining factors driving incorporation of land use approaches into hazard mitigation plans. Journal of Environmental Planning and Management 57(5): 792-811.

Malanson, G. P. 2002. Extinction-debt trajectories and spatial patterns of habitat destruction. Annals of the Association of American Geographers 92: 177-188.

Manning, L. M., C. H. Peterson, and S. R. Fegley. 2013. Degradation of surf-fish foraging habitat driven by persistent sedimentological modifications caused by beach nourishment. Bulletin of Marine Science 89: 83-106.

Marcomini, S. C., and R. A. López. 2006. Evolution of a beach nourishment project at Mar del Plata. Journal of Coastal Research SI39: 834-837.

Martínez, M. L., and J. G. García-Franco. 2004. Plant-plant interactions in coastal dunes. Pp. 205-220 in Coastal Dunes, Ecology and Conservation, edited by M. L. Martínez and N. P. Psuty. Berlin: Springer-Verlag.

Maslo, B., S. N. Handel, and T. Pover. 2011. Restoring beaches for Atlantic Coast piping plovers (*Charadrius melodus*): A classification and tree analysis of nest-site selection. Restoration Ecology 19: 194-203.

Maun, M. A. 2004. Burial of plants as a selective force in sand dunes. Pp. 119-135 in Coastal Dunes, Ecology and Conservation, edited by M. L. Martínez and N. P. Psuty. Berlin: Springer-Verlag.

Mauriello, M. N. 1991. Beach nourishment and dredging: New Jersey's policies. Shore and Beach 59(3): 25-28.

May, P., R. Burby, N. Ericksen, J. Handmer, J. Dixon, S. Michaels, and D. I. Smith. 1996. Environmental Management and Governance: Intergovernmental Approaches to Hazards and Sustainability. London: Routledge Press.

Mazda, Y., M. Magi, Y. Ikeda, T. Kurokawa, and T. Asano. 2006. Wave reduction in a mangrove forest dominated by *Sonneratia* sp. Wetlands Ecology and Management 14(4): 365-378.

McBride, R. A., M. R. Byrnes, and H. W. Hiland. 1995. Geomorphic response-type model for barrier coastlines: A regional perspective. Marine Geology 126(1): 143-159.

McIvor, A. L., T. Spencer, I. Möller, and M. Spalding. 2012. Storm Surge Reduction by Mangroves. Natural Coastal Protection Series, Report 2. Cambridge Coastal Research Unit Working Paper 41. The Nature Conservancy and Wetlands International. Available at http://coastalresilience.org/sites/default/files/resources/storm-surge-reduction-by-mangroves-report.pdf.

Mcleod, E., G. L. Chmura, S. Bouillon, R. Salm, M. Björk, C. M. Duarte, C. E. Lovelock, W. H. Schlessinger, and B. R. Silliman. 2011. A blueprint for blue carbon: Toward an improved understanding of the role of vegetated coastal habitats in sequestering CO_2. Frontiers in Ecology and the Environment 9(10): 552-560.

Melillo, J. M., T. C. Richmond, and G. W. Yohe (eds.). 2014. Climate Change Impacts in the United States: The Third National Climate Assessment. U.S. Global Change Research Program. doi:10.7930/J0Z31WJ2.

Mendelssohn, I. A., M. W. Hester, F. J. Monteferrante, and F. Talbot. 1991. Experimental dune building and vegetative stabilization in a sand-deficient barrier island setting on the Louisiana coast, USA. Journal of Coastal Research 7: 137-149.

Michel, J., R. Nairn, C. H. Peterson, S. W. Ross, R. Weisberg, and R. Randall. 2007. Critical Technical Review and Evaluation of Site-Specific Studies Techniques for the MMS Marine Minerals Program. OCS Study MMS 2007-047. U.S. Department of the Interior, Marine Minerals Service. Available at http://www.boem.gov/Non-Energy-Minerals/2007-047.aspx.

Michel-Kerjan, E. 2010. Catastrophe economics: The National Flood Insurance Program. Journal of Economic Perspectives 24: 165-186.

Michel-Kerjan, E. O. 2013. Have We Entered an Ever-Growing Cycle on Government Disaster Relief? U.S. Senate Committee on Small Business and Entrepreneurship, Washington, DC. Available at http://opim.wharton.upenn.edu/risk/library/US-Senate-Small-Business-Cte_2013Mar14_MichelKerjan.pdf.

Mileti, D. 1999. Disasters by Design: A Reassessment of Natural Hazards in the United States. Washington, DC: Joseph Henry Press.

Minello, T. J., R. J. Zimmerman, and R. Medina. 1994. The importance of edge for natant macrofauna in a created salt-marsh. Wetlands 14(3): 184-198.

Minerals Management Service. 2001. Development and Design of Biological and Physical Monitoring Protocols to Evaluate the Long-Term Impacts of Offshore Dredging Operations on the Marine Environment. Final Report MMS 2001-089. U.S. Department of the Interior, Minerals Management Service. Available at http://www.boem.gov/Non-Energy-Minerals/2001-089.aspx.

Ministerie van Verkeer en Waterstaat [Ministry of Transport]. 2007. Hydraulisch Randvoorwaardenboek 2006 [Hydraulic Conditions for Dune 2006 Flood] (in Dutch).

Mishan, E. 1971. Evaluation of life and limb. Journal of Political Economy 79(4): 687-705.

Moore, J. W., and D. P. Moore. 1989. The Army Corps of Engineers and the Evolution of Federal Flood Plain Management Policy. Boulder Institute of Behavioral Science, University of Colorado.

Morang, A. 2007. Hurricane Barriers in New England and New Jersey—History and Status After Four Decades. ERDC/CHL TR-07-11. U.S. Army Engineer District, New Orleans. Available at http://www.dtic.mil/dtic/tr/fulltext/u2/a473784.pdf.

Moschella, P. S., M. Abbiati, P. Aberg, L. Airoldi, J. M. Anderson, F. Bacchiocchi, F. Bulleri, G. E. Dinesen, M. Frost, E. Gacia, L. Granhag, P. R. Jonsson, M. P. Satta, A. Sundelöf, R. C. Thompson, and S. J. Hawkins. 2005. Low-crested coastal defense structures as artificial habitats for marine life: Using ecological criteria in design. Coastal Engineering 52: 1053-1071.

Mrozek, J., and L. O. Taylor. 2002. What determines the value of life? A meta-analysis. Journal of Policy Analysis and Management 21(2): 253-270.

Murray, C., and D. R. Marmorek. 2004. Adaptive management: A science-based approach to managing ecosystems in the face of uncertainty. In Making Ecosystem-Based Management Work: Proceedings of the Fifth International Conference on Science and Management of Protected Areas, Victoria, British Columbia, edited by N. W. P. Munro, T. B. Herman, K. Beazley, and P. Dearden. Wolfville, Nova Scotia: Science and Management of Protected Areas Association.

Nagelkerken, I., S. Bouillon, P. Green, and M. Haywood. 2008.The habitat function of mangroves for terrestrial and marine fauna: A review. Aquatic Botany 89(2): 155-185.

Nelson, A. C., and S. P. French. 2002. Plan Quality and Mitigating Damage from Natural Disasters: A Case Study of the Northridge Earthquake with Planning Policy Considerations. Journal of the American Planning Association 68: 194-207.

Nelson, W. G. 1989. Beach nourishment and hard bottom habitats: The case for caution. Pp. 109-116 in Proceedings of the 1989 National Conference on Beach Preservation Technology. Tallahassee: Florida Shore and Beach Preservation Association.

Nelson, W.G. 1993. Beach restoration in the southeastern US: Environmental effects and biological monitoring. Ocean and Coastal Management 19: 157-182.

Nepf, H.M., 1999. Drag, turbulence, and diffusion in flow through emergent vegetation. Water Resources Research (35) 2: 479-489.

New Orleans City Planning Commission. 1999. New Century New Orleans, 1999 Land Use Plan, City of New Orleans. New Orleans City Planning Commission.

Newell, R. C., L. J. Seiderer, N. M. Simpson, and J. E. Robinson. 2004. Impacts of marine aggregate dredging on benthic macrofauna of the south coast of the United Kingdom. Journal of Coastal Research 20: 115-125.

NOAA (National Oceanic and Atmospheric Administration).1979. Meteorological criteria for standard project hurricane and probable maximum hurricane windfields, gulf and east coasts of the United States. NOAA Technical Report.

NOAA. 2007. Beach Noursihment: A Guide for Local Government Officials. Available at http://www.csc.noaa.gov/beachnourishment/html/human/law/sec404.htm.

NOAA. 2013. FY2014 Budget Summary. Available at http://www.corporateservices.noaa.gov/nbo/fy14_bluebook/FINALnoaaBlueBook_2014_Web_Full.pdf.

Nordhaus, W. D. 2007. A review of the Stern Review of the Economics of Climate Change. Journal of Economic Literature 45: 686-702.

Nordstrom, K. F. 2005. Beach nourishment and coastal habitats: Research needs for improving compatibility. Restoration Ecology 13: 215-222.

Nordstrom, K. F. 2008. Beach and Dune Restoration. Cambridge, UK. Cambridge University Press.

Nordstrom, K. F. In press. Living with shore protection structures: A review. Estuarine Coastal and Shelf Science, doi: 10.1016/j.ecss.2013.11.003.

Nordstrom, K. F., R. Lampe, and L. M. Vandemark. 2000. Re-establishing naturally-functioning dunes on developed coasts. Environmental Management 25: 37-51.

Nordstrom, K. F., N. L. Jackson, N. C. Kraus, T. W. Kana, R. Bearce, L. M. Bocamazo, D. R. Young, and H. A. DeButts. 2011. Enhancing geomorphic and biologic functions and values on backshores and dunes of developed shores: A review of opportunities and constraints. Environmental Conservation 38: 288-302.

Nordstrom, K. F., N. L. Jackson, A. L. Freestone, K. H. Korotky and J. A. Puleo. 2012. Effects of beach raking and sand fences on dune dimensions and morphology. Geomorphology 179: 106-115.

NPCC (New York City Panel on Climate Change). 2013. Climate Risk Information 2013: Observations, Climate Change Projections, and Maps. C. Rosenzweig and W. Solecki (eds.). Prepared for use by the City of New York Special Initiative on Rebuilding and Resilency, New York.

NRC (National Research Council). 1994. Science and Judgment in Risk Assessment. Washington, DC: National Academy Press.
NRC. 1995. Beach Nourishment and Protection. Washington, DC: National Academy Press.
NRC. 1996. Understanding Risk: Informing Decisions in a Democratic Society, edited by P. Stern and H. Fineberg. Washington, DC: National Academy Press.
NRC. 1999. New Directions in Water Resources Planning for the U.S. Army Corps of Engineers. Washington, DC: National Academy Press.
NRC. 2001. Building Community Disaster resilience Through Private-Public Collaboration. Washington, DC: National Academy Press.
NRC. 2004a. Adaptive Management for Water Resources Project Planning. Washington, DC: The National Academies Press.
NRC. 2004b. Analytical Methods and Approaches for Water Resources Project Planning. Washington, DC: The National Academies Press.
NRC. 2007. Migitating Shore Erosion Along Sheltered Coasts. Washington, DC: The National Academies Press.
NRC. 2009. Mapping the Zone: Improving Flood Map Accuracy. Washington, DC: The National Academies Press.
NRC. 2010. Private-Public Sector Collaboration to Enhance Community Resilience: A Workshop Report. Washington, DC: National Academies Press.
NRC. 2011a. Assessing National Resilience to Hazards and Disasters: The Perspective from the Gulf Cost of Louisiana and Mississippi: Summary of a Workshop. Washington, DC: The National Academies Press.
NRC. 2011b. National Water Resources Challenges Facing the U.S. Army Corps of Engineers. Washington, DC: The National Academies Press.
NRC. 2012a. Corps of Engineers Water Resources Infrastructure: Deterioration, Investment, or Divestment? Washington, DC: National Academies Press.
NRC. 2012b. Disaster Resilience A National Imperative. Washington, DC: The National Academies Press.
NRC. 2012c. Sea-Level Rise for the Coasts of California, Oregon, and Washington: Past, Present, and Future. Washington, DC: The National Academies Press.
NRC. 2013. Levees and the National Flood Insurance Program: Improving Policies and Practice. Washington, DC: The National Academies Press.
NSW-DSC (New South Wales Government, Dams Safety Committee). 2006. Risk Management Policy Framework for Dam Safety. Parramatta: New South Wales Government, Dams Safety Committee.
NYC (New York City). 2013. A Stronger More Resilient New York. Available at http://www.nyc.gov/html/sirr/html/report/report.shtml.
OMB (Office of Management and Budget). 2003. Regulatory Analysis. Circular A-4. Executive Office of the President, Office of Management and the Budget, Washington, DC.
OMB. 2010. Report to Congress on the Benefits and Costs of Federal Regulations and Unfunded Mandates on State, Local, and Tribal Entities. Washington, DC: Office of Management and the Budget, and Office of Technology Policy.
Olshansky, R. B. 2001. Land Use Planning for Seismic Safety: The Los Angeles County Experience, 1971-1994. Journal of the American Planning Association 67: 173-185.
Ostrom, E., 2010, Polycentric systems for coping with collective action and global environmental change. Global Environmental Change 20: 550-557.
Ousley, J. D., E. Kromhout, and M. H. Schrader. 2012. Southeast Florida Sediment Assessment and Needs Determinations (SAND) Study. U.S. Army Corps of Engineers Jacksonville District.

Ozeren, Y., and D. G. Wren. 2010. Laboratory measurements of wave attenuation through model and live vegetation. Pp. 45-56 in Coastal Hazards, edited by W. Huang, K.-H. Wang, and Q. J. Chen. Reston, VA: American Society of Civil Engineers.

Painter, W., and J. T. Brown. 2013. FY2013 Supplemental Funding for Disaster Relief. Congressional Research Service R42869.

Peterson, C. H., and M. J. Bishop. 2005. Assessing the environmental impacts of beach nourishment. Bioscience 55: 887-896.

Peterson, C. H., M. J. Bishop, G. A. Johnson, L. M. D'Anna, and L. M. Manning. 2006. Exploiting beach filling as an unaffordable experiment: Benthic intertidal impacts propagating up to shorebirds. Journal of Experimental Marine Biology and Ecology 338: 205-221.

Peterson, C. H., M. J. Bishop, L. M. D'Anna, and G. A. Johnson. 2014. Multi-year persistence of beach habitat degradation from nourishment using coarse shelly sediments. Science of the Total Environment 487: 481-492.

Pezzuto, P. R., C. Resgalla, Jr., J. G. N. Abreu, and J. T. Menezes. 2006. Environmental impacts of the nourishment of Balneário Camboriú beach, SC, Brazil. Journal of Coastal Research SI39: 863-868.

Pielke, R. A., Jr., J. Gratz, C. W. Landsea, D. Collins, M. A. Saunders, and R. Musulin. 2008. Normalized hurricane damages in the United States: 1900-2005. Natural Hazards Review 9(1): 29-42.

Pilkey, O. H. 1992. Another view of beachfill performance. Shore and Beach 60(2): 20-25.

Pilkey, O. H., and H. L. Wright. 1988. Seawalls versus beaches. Journal of Coastal Research SI4: 41-66.

Pimm, S. L., and P. Raven. 2000. Extinction by numbers. Nature 403: 843-845.

Plant, N. G., and G. B. Griggs. 1992. Interactions between nearshore processes and beach morphology near a seawall. Journal of Coastal Research 8:183-200.

Platt, R. H., D. Salvesen, and G. H. Baldwin. 2002. Rebuilding the North Carolina coast after Hurricane Fran: Did public regulations matter? Coastal Management 30: 249-269.

Polasky, S., and S. Binder. 2012. Valuing the environment for decision-making. Issues in Science and Technology 28(4): 53-62.

Porter, K. 2002. Life safety risk criteria in seismic decisions. Pp. 135-159 in Acceptable Risk Processes: Lifelines and Natural Hazards, edited by C. Taylor and E. H. VanMarcke. New York: American Society of Civil Engineers.

Posey, M., and T. Alphin. 2002. Resilience and stability in an offshore benthic community: Responses to sediment borrow activities and hurricane disturbance. Journal of Coastal Research 18: 685-697.

Powers, S. P., and K. E. Boyer. 2014. Marine restoration ecology. Chapter 22 in Marine Community Ecology and Conservation, edited by M. D. Bertness, J. F. Bruno, B. R. Silliman, and J. J. Stachowicz. Sunderland, MA: Sinauer Associates.

Powers, S. P., C. H. Peterson, J. H. Grabowski, and H. S. Lenihan. 2009. Evaluating the success of constructed oyster reefs in no-harvest sanctuaries: Implications for restoration. Marine Ecology Progress Series 389: 159-170.

Quartel, S., A. Kroon, P. Augustinus, P. Van Santen, and N. H. Tri. 2007. Wave attenuation in coastal mangroves in the Red River Delta, Vietnam. Journal of Asian Earth Sciences 29(4): 576-584.

Quay, R. 2010. Anticipatory Governance: A Tool for Climate Change Adaptation. Journal of the American Planning Association 76(4): 496-511.

Rakocinski, C. F., R. W. Heard, S. E. LeCroy, J. A. McLelland, and T. Simons. 1996. Responses by macrobenthic assemblages to extensive beach restoration at Perdido Key, Florida, U.S.A. Journal of Coastal Research 12: 326-353.

RAND Corp. 2012. Financing the Operation and Maintenance Costs of Hurricane Protection Infrastructure: Options for the State of Louisiana. TR-1223-OCPR. Santa Monica, CA: RAND Corp. Available at http://www.rand.org/content/dam/rand/pubs/technical_reports/2012/RAND_TR1223.pdf.

RECOVER (Restoration Coordination & Verification). 2011. The Comprehensive Everglades Restoration Plan Adaptive Management Integration Guide. Restoration Coordination and Verification. U.S. Army Corps of Engineers, Jacksonville District, and South Florida Water Management District, West Palm Beach, FL.

Reilly, F. J., and V. J. Bellis. 1983. The ecological impact of beach nourishment with dredged materials on the intertidal zone at Bogue Banks, North Carolina. Miscellaneous Report 83-3. Ft. Belvoir, VA: U.S. Army Corps of Engineers, Coastal Engineering Research Center.

Renaud, F. G., K. Sudmeier-Rieux, and M. Estrella (eds). 2013. The Role of Ecosystems in Disaster Risk Reduction. New York: United Nations University Press.

Resio, D. T., and J. J. Westerink. 2008. Modeling the physics of storm surges. Physics Today 61: 33-38.

Robinson, M. F. 2004. History of the 1% Chance Flood Standard. Pp. 2-8 in Reducing Flood Losses: Is the 1% Chance (100-Year) Flood Standard Sufficient? Assembly of the Gilbert F. White National Flood Policy Forum. Available at ftp://ftp-fc.sc.egov.usda.gov/Economics/Technotes/ASFFM-1in100Chance-FPpapers.pdf [accessed June 18, 2014].

Rodriguez, A. B., F. J. Fodrie, J. T. Ridge, N. L. Lindquist, E. J. Theuerkauf, S. E. Coleman, J. H. Grabowski, M. C. Brodeur, R. K. Gittman, D. A. Keller, and M. D. Kenworthy. 2014. Oyster reefs can outpace sea-level rise. Nature Climate Change 4: 493-497.

Rogers, S. M. 2007. Beach nourishment for hurricane protection: North Carolina project performance in Hurricanes Dennis and Floyd. Shore and Beach 75(1).

Roscoe, K. L., and F. Diermanse. 2011. Effect of surge uncertainty on probabilistically computed dune erosion. Coastal Engineering 58(11): 1023-1033.

Rose, A., K. Porter, N. Dash, J. Bouabid, C. Huyck, J. Whitehead, D. Shaw, R. Eguchi, C. Taylor, T. McLane L. T.Tobin, P. T. Ganderton, D. Godschalk, A/S. Kiremidjian, K. Tierney, and C. T. West. 2007. Benefit-cost analysis of FEMA hazard mitigation grants. Natural Hazards Review 8(4): 97-111.

Saeijs, H. L. F., and A. J. M. Geurts van Kessel. 2005. The Oosterschelde, a changing ecosystem after completion of the delta works. Pp 317-334 in Flooding and Environmental Challenges for Venice and Its Lagoon: State of Knowledge, edited by C. A. Fletcher and T. Spencer. Cambridge, UK: Cambridge University Press.

Sallenger, A. H. 2000. Storm impact scale for barrier islands. Journal of Coastal Research 16: 890-895.

Sallenger, A. H., Jr., K. S. Doran, and P. A. Howd. 2012. Hotspot of accelerated sea-level rise on the Atlantic coast of North America. Nature Climate Change 2: 884-888.

Salvesen, D. 2005. The Coastal Barrier Resources Act: Has it discouraged coastal development? Coastal Management 33: 181-195.

Sagoff, M. 1988. The economy of the Earth: Philosophy, law, and the environment. Cambridge, UK: Cambridge University Press.

Schlacher, T. A., J. Dugan, D. S. Schoeman, M. Lastra, A. Jones, F. Scapini, A. McLachlan, and O. Defeo. 2007. Sandy beaches at the brink. Diversity and Distributions 13: 556-560.

Schlacher, T. A., R. Noriega, A. Jones, and T. Dye. 2012. The effects of beach nourishment on benthic invertebrates in eastern Australia: Impacts and variable recovery. Science of the Total Environment 435: 411-417.

Schupp, C. A., N. T. Winn, T. L. Pearl, J. P. Kumer, T. J. B. Carruthers, and C. S. Zimmerman. 2013. Restoration of overwash processes creates piping plover (*Charadrius melodus*) habitat on a barrier island (Assateague Island, Maryland). Estuarine Coastal and Shelf Science 116: 11-20.

Scyphers, S. B., S. P. Powers, K. L. Heck, Jr., and D. Byron. 2011. Oyster reefs as natural breakwaters mitigate shoreline loss and facilitate fisheries. PLoS ONE 6: e22396.

Seabloom, E. W., A. P. Dobson, and D. M. Stoms. 2002. Extinction rates under nonrandom patterns of habitat loss. Proceedings of the National Academy of Sciences of the United States of America 99: 11229-11234.

SFRCCC (Southeast Florida Regional Climate Change Compact). 2012. A Region Responds to a Changing Climate: Southeast Florida Regional Climate Action Plan. Available at http://southeastfloridaclimatecompact.org/pdf/Regional Climate Action Plan FINAL ADA Compliant.pdf.

Sheng, Y. P., A. Lapetina, and G. Ma. In press. The reduction of storm surge by vegetation canopies: Three-dimensional simulations. Geophysical Research Letters, doi:10.1029/2012GL053577.

Shepard, C., C. M. Crain, and M. W. Beck. 2011. The protective role of coastal marshes: A systematic review and meta-analysis. PLoS ONE 6(11): e27374.

Shipman, H. 2001. Beach nourishment on Puget Sound: A review of existing projects and potential applications. In Puget Sound Research 2001. Olympia, WA: Puget Sound Water Quality Action Team.

Slovic, P. 1987. Perception of risk. Science 236: 280-285.

Slovic, P. 2000. The Perception of Risk. Florence, KY: Routledge.

Smith, V. K., and J.-C. Huang. 1995. Can markets value air quality? A meta-analysis of hedonic property value models. Journal of Political Economy 103: 209-227.

Smith, A. B., and R. W. Katz. 2013. U.S. billion-dollar weather and climate disasters: Data sources, trends, accuracy and biases. Natural Hazards 67(2): 387-410.

Sobocinski, K. L., J. R. Cordell, and C. A. Simenstad. 2010. Effects of shoreline modifications on supratidal macroinvertebrate fauna on Puget Sound, Washington beaches. Estuaries and Coasts 33: 699-711.

Speybroeck, J., D. Bonte, W. Courtens, T. Gheskiere, P. Grootaert, J.-P. Maelfait, M. Mathys, S. Provoost, K. Sabbe, E. W. M. Steinen, V. van Lancker, M. Vincx, and S. Degraer. 2006. Beach nourishment: An ecologically sound coastal defense alternative? A review. Aquatic Conservation: Marine and Freshwater Ecosystems 16: 419-435.

Starr, C. 1969. Social benefit versus technological risk. Science 165: 1232-1238.

Staszak, L. A., and A. R. Armitage. 2012. Evaluating salt marsh restoration success with an Index of Ecosystem Integrity. Journal of Coastal Research 29(2): 410-418.

Stauble, D. K., W. C. Seabergh, and L. Z. Hales. 1991. Effects of Hurricane Hugo on the South Carolina coast. Journal of Coastal Research SI8: 129-162.

Stern, N. 2007. The Economics of Climate Change: The Stern Review. Cambridge, UK, and New York: Cambridge University Press.

Stern, N. 2008. The economics of climate change. American Economic Review 98(2): 1-37.

Sweet, W., C. Zervas, S. Gill, and J. Park. 2013. Hurricane Sandy inundation probabilities today and tomorrow. Bulletin of the American Meteorological Society 94(9): S17-S20.

Sylves, R., and Z. I. Buzas. 2007. Presidential disaster declaration decisions, 1953-2003: What influences the odds? State and Local Government Review 39(1): 3-15.

Tanaka, K. 2008. Effectiveness and limitation of the coastal vegetation for storm surge disaster mitigation. Pp. 60-73 in Investigation Report on the Storm Surge Disaster by Cyclone Sidr in 2007, Bangladesh. Investigation Team of Japan Society of Civil Engineering. Available at http://www.jsce.or.jp/report/46/files/Bangladesh_Investigation.pdf.

Teal, J., and S. Peterson. 2009. The use of science in the restoration of northeastern U.S. salt marshes. Pp. 267-284 in Human Impacts on Salt Marshes: A Global Perspective, edited by B. R. Silliman, E. Grosholz, and M. D. Bertness. Berkeley: University of California Press.

Temmerman, S., P. Meire, T. J. Bouma, P. M. J. Herman, T. Ysebaert, and H. J. De Vriend. 2013. Ecosystem-based coastal defence in the face of global change. Nature 504: 79-83.

Thieler, E. R., and R. S. Young. 1991. Quantitative evaluation of coastal geomorphological changes in South Carolina after Hurricane Hugo. Journal of Coastal Research SI8: 187-200.

Tierney, K., and M. Bruneau. 2007. Conceptualizing and measuring resilience: A key to disaster loss reduction. TR News 250(May-June): 14-18.

Tierney, K., M. Lindell, and R. Perry. 2001. Facing the Unexpected: Disaster Preparedness and Response in the United States. Washington, DC: Joseph Henry Press.

Titus, J. G., K. E. Anderson, D. R. Cahoon, D. B. Gesch, S. K. Gill, B. T. Gutierrez, E. R. Thieler, and S. J. Williams. 2009. Coastal Sensitivity to Sea-Level Rise: A Focus on the Mid-Atlantic Region. A report by the U.S. Climate Change Science Program and the Subcommittee on Global Change Research. Washington, DC: U.S. Environmental Protection Agency. Available at http://downloads.globalchange.gov/sap/sap4-1/sap4-1-final-report-all.pdf.

UNISDR (United Nations Office for Disaster Risk Reduction). 2009. Terminology. Available at http://www.unisdr.org/we/inform/terminology [accessed July 2, 2014].

UNISDR. 2014. Disaster Resilience Scorecard for Cities. Geneva: UN Office for Disaster Risk Reduction.

USACE (U.S. Army Corps of Engineers). 1962. Coastal Storm of 6-7 March 1962 Post Flood Report. USACE Philadelphia District.

USACE 1963. Report on Operation Five-High: March 1962 Storm. U.S. Army Corps of Engineers, North Atlantic Division.

USACE. 1965. Interim Hurricane Study of Lake Pontchartrain and Vicinity, Louisiana. Washington, DC: U.S. Government Printing Office.

USACE. 1981.Galveston's Bulkwark Against the Sea: History of the Galveston Seawall. U.S. Army Corps of Engineers, Galveston District.

USACE. 1996. Final Report: An Analysis of the U.S. Army Corps of Engineers Shore Protection Program. IWR Report 96-PS-1. Alexandria, VA Institute for Water Resources, U.S. Army Corps of Engineers Water Resources Support Center. Available at http://www.iwr.usace.army.mil/Portals/70/docs/iwrreports/96-PS-1.pdf.

USACE. 2000a. Hurricane Fran Effects on Communities with and Without Shore Protection: A Case Study at Six North Carolina Beaches. IWR Report 00-R-6.

USACE. 2000b. Planning Guidance Notebook. ER 1105-2-100, April 22, 2000. Available at http://planning.usace.army.mil/toolbox/library/ERs/entire.pdf [accessed February 11, 2014].

USACE. 2001. The New York District's Biological Monitoring Program for the Atlantic Coast of New Jersey, Asbury Park to Manasquan Section Beach Erosion Control Project. Engineer Research and Development Center, Waterways Experiment Station, Vicksburg, MS.

USACE. 2009. Mississippi Coastal Improvements Program (MsCIP), Hancock, Harrison, and Jackson Counties, Mississippi: Comprehensive Plan and Integrated Programmatic Environmental Impact Statement. Volume 1, Main Report. Available at http://www.sam.usace.army.mil/Portals/46/docs/program_management/mscip/docs/MSCIP%20Main%20Report%20062209-Errata.pdf.

USACE. 2011a. Coastal Storm Risk Management: National Economic Development Manual. IWR Report 2011-R-09. Available at http://www.iwr.usace.army.mil/Portals/70/docs/iwrreports/2011-R-09.pdf.

USACE. 2011b. Sea Level Change Considerations for Civil Works Programs. EC 1165-2-212. Available at http://planning.usace.army.mil/toolbox/library/ECs/EC11652212Nov2011.pdf.

USACE. 2012a. Hurricane Sandy aftermath: Hurricane barriers managed by Corps engineers in New England prevent $29.7 million in damages. Press Release, November 21. Available at http://www.nae.usace.army.mil/Media/NewsReleases/tabid/11736/Article/10404/hurricane-sandy-aftermath-hurricane-barriers-managed-by-corps-engineers-in-new.aspx.

USACE. 2012b. Greater New Orleans Hurricane Storm Damage Risk Reduction System: Facts and Figures. August 2012. Available at: http://www.mvn.usace.army.mil/Portals/56/docs/PAO/Brochures/FactsFiguresAugust 2012.pdf.

USACE. 2013a. Coastal Risk Reduction and Resilience. CWTS 2013-3. Washington, DC: Directorate of Civil Works, U.S. Army Corps of Engineers.

USACE. 2013b. Greater New Orleans Hurricane and Storm Damage Risk Reduction System: Facts and Figures. June 2013. Available at http://www.mvn.usace.army.mil/Portals/56/docs/HSDRRS/Facts-Figures2013.pdf.

USACE. 2013c. Hurricane Issac With and Without 2012 100-Year HSDRSS Evaluation. Final Report. Available at http://www.mvn.usace.army.mil/Portals/56/docs/PAO/20130208 HurrIsaacW-WO2012HSDRRS.pdf.

USACE. 2013d. Hurricane Sandy Coastal Projects Performance Evaluation Study: Disaster Relief Appropriations Act, 2013. Available at http://www.nan.usace.army.mil/Portals/37/docs/civilworks/SandyFiles/USACE_Post-Sandy_Coastal_Projects_Performance_Evaluation_Study.pdf.

USACE. 2013e. Incorporating Sea Level Change in Civil Works Programs. Engineering Regulation 1100-2-8162. Washington, DC: Department of the Army.

USACE. 2013f. North Atlantic Coast Comprehensive Study: Resilient Adaptation to Increasing Risk—HQ Review Draft. December 20. Brooklyn, NY: U.S. Army Corps of Engineers, North Atlantic Division.

USACE. 2013g. Corps of Engineers Civil Works Direct Program Budget Development Guidance, Fiscal Year 2015. EC 11-2-204.

USACE. 2014. Flagler County, Florida: Hurricane and Storm Damage Reduction Project, Draft Integrated Feasibility Study and Environmental Assessment. U.S. Army Corps of Engineers, Jacsonville District. Available at http://www.saj.usace.army.mil/Portals/44/docs/Planning/EnvironmentalBranch/EnvironmentalDocs/FlaglerCoSPP_MainJan2014.pdf.

USBR (U.S. Bureau of Reclamation). 2003. Guidelines for Achieving Public Protection in Dam Safety Decision Making. Denver: USBR. Available at http://www.usbr.gov/ssle/damsafety/Risk/ppg2003.pdf.

U.S. Commission on Ocean Policy. 2004. An Ocean Blueprint for the 21st Century: Final Report. Washington, DC: U.S. Commission on Ocean Policy. Available at http://www.opc.ca.gov/webmaster/ftp/pdf/docs/Documents_Page/Reports/U.S.%20Ocean%20Comm%20Report/FinalReport.pdf.

U.S. Inter-Agency River Basin Committee Subcommittee on Benefits and Costs. 1950. Report to the Federal Inter-Agency River Basin Committee: Proposed Practices for Economic Analysis of River Basin Projects. Washington, DC.

van der Most, H., and M. Wehrung. 2005. Dealing with uncertainty in flood risk assessment of dike rings in the Netherlands. Natural Hazards 36: 191-206.

van der Wal, D. 1998. The impact of the grain-size distribution of nourishment sand on aeolian sand transport. Journal of Coastal Research 14: 620-631.

Van Tomme, J., S. Vanden Eede, J. Speybroeck, S. Degraer, and M. Vincx. 2012. Macrofaunal sediment selectivity considerations for beach nourishment programmes. Marine Environmental Research 84: 10-16.

Vecchi, G. A., and T. R. Knutson. 2011. Estimating Annual Numbers of Atlantic Hurricanes Missing from the HURDAT Database (1878-1965) Using Ship Track Density. Journal of Climate 24: 1736-1746.
VIMS (Virginia Institute of Marine Sciences). 2013. Recurrent Flooding Study for Tidewater Virginia. Available at http://ccrm.vims.edu/recurrent_flooding/Recurrent_Flooding_Study_web.pdf.
Viola, S., J. E. Dugan, D. M. Hubbard, and N. K. Schooler. In press. Burrowing inhibition by fine-textured beach fill: Implications for recovery of beach ecosystems. Estuarine, Coastal and Shelf Science.
Viscusi, W. K. 2004. The value of life: Estimates with risks by occupation and industry. Economic Inquiry 42(1): 29-48.
Viscusi, W. K., and J. E. Aldy. 2003. The value of a statistical life: A critical review of market estimates throughout the world. Journal of Risk and Uncertainty 27(1): 5-76.
Vose, F. E. and S. S. Bell. 1994. Resident fishes and macrobenthos in mangrove rimmed habitats: evaluation of habitat restoration by hydrologic modification. Estuaries 17:585-596.
Vrijling, J. K., W. van Hengel, and R. J. Houben. 1995. A framework for risk evaluation. Journal of Hazardous Materials 43(3): 245-261.
Vrijling, J., W. van Hengel, and R. Houben. 1998. Acceptable risk as a basis for design. Reliability Engineering and System Safety 59: 141-150.
Walsh, M. J. 2012. U.S. Army Corps of Engineers Civil Works Feasibility Study Program Execution and Delivery. Memorandum for Major Subordinate Commands, February 8. Available at http://planning.usace.army.mil/toolbox/library/MemosandLetters/USACE_CW_FeasibilityStudyProgramExecutionDelivery.pdf.
Walsh, S., and R. Miskewitz. 2013. Impact of sea level rise on tide gate function. Journal of Environmental, Science and Health, Part A 48: 453-463.
Walters, C. 1986. Adaptive Management of Renewable Resources. New York: Macmillan.
Wamsley, T. V., M. A. Cialone, J. M. Smith, J. H. Atkinson, and J. D. Rosati. 2010. The potential of wetlands in reducing storm surge. Ocean Engineering 37(1): 59-68.
Wang, P., K. E. Brutsche, T. M. Beck, J. D. Rosati, and L. S. Lillycrop. 2013. Initial Morphological Evolution of Perdido Key Berm Nourishment, Florida. Coastal and Hydraulics Engineering Technical Note ERDC/CHL CHETN-IV-89. Vicksburg, MS: US Army Engineer Research and Development Center, Coastal and Hydraulics Laboratory. Available at http://chl.erdc.usace.army.mil/library/publications/ chetn/pdf/chetn-xiv-?.pdf.
Weggel, J. R. 1988. Seawalls: The need for research, dimensional considerations and a suggested classification. Journal of Coastal Research SI4: 29-40.
Weitzman, M. L. 2001. Gamma Discounting. American Economic Review 91(1): 260-271.
Wiegel, R. L., and T. Saville. 1996. History of Coastal Engineering in the US. Pp. 513-600 in History and Heritage of Coastal Engineering, edited by N. C. Kraus. American Society of Civil Engineers.
Williams, S. J. 1986. Sand and gravel deposits within the United States exclusive economic zone: Resource assessment and uses. Pp. 377-386 in Proceedings of the 18th Offshore Technology Conference. Houston, TX: Offshore Technology Conference.
Williams, S. J., and M. K. Foley. 2007. Recommendations for a Barrier Island Breach Management Plan for Fire Island National Seashore, Including the Otis Pike High Dune Wilderness Area, Long Island, New York. Technical Report NPS/NER/NRTR—2007/075. Boston, MA: National Park Service. Available at http://www.nps.gov/fiis/naturescience/upload/2007_BreachManagementRecommendations_FIIS%20NRTR-2007-075.pdf.
Williams, S. J., J. Flocks, C. Jenkins, S. Khalil, and J. Moya. 2012. Offshore sediment character and sand resource assessment of the northern Gulf of Mexico, Florida to Texas. Journal of Coastal Research: SI60:30-44.

Woodhouse, W. W., Jr., E. D. Seneca, and S. W. Broome. 1977. Effect of species on dune grass growth. International Journal of Biometeorology 21: 256-266.

Woolley, D., and L. Shabman. 2008. Decision-Making Chronology for the Lake Pontchartrain and Vicinity Hurricane Protection Project. Final Report for the Headquarters, U.S. Army Corps of Engineers. Available at http://biotech.law.lsu.edu/katrina/hpdc/Final_HPDC_Apr3_2008.pdf.

WRC (U.S. Water Resources Council). 1973. Water and related land resources: Establishment of principles and standards for planning. Federal Register 38: 24778. Available at http://planning.usace.army.mil/toolbox/library/Guidance/water_relatedladres.pdf.

WRC. 1983. Economic and Environmental Principles and Guidelines for Water and Related Implementation Studies. Washington, DC: U.S. Government Printing Office. Available at http://planning.usace.army.mil/toolbox/library/Guidance/Principles_Guidelines.pdf.

Wright, J. M. 2000. The Nation's Responses to Flood Disasters: A Historical Account. Madison, WI: Association of State Floodplain Managers. Available at http://www.floods.org/PDF/hist_fpm.pdf.

Younger, M., H. R. Morrow-Almeida, S. M. Vindigni, and A. L. Dannenberg. 2008. The built environment, climate change, and health: opportunities for co-benefits. American Journal of Preventive Medicine 35:517-526.

Zelo, I., H. Shipman, and Brennan. 2000. Alternative bank protection methods for Puget Sound shorelines. Ecology Publication No. 00-06-012. Olympia: Washington Department of Ecology.

Zhang, K. Q., H. Liu, Y. Li, X. Hongzhou, S. Jian, J. Rhome, and T. J. Smith III. 2012. The role of mangroves in attenuating storm surges. Estuarine, Coastal and Shelf Science 102: 11-23.

Zu Ermgassen, P. S. E., M. D. Spalding, B. Blake, L. D. Coen, B. Dumbauld, S. Geiger, J. H. Grabowski, R. Grizzle, M. Luckenbach, K. McGraw, W. Rodney, J. L. Ruesink, S. P. Powers, and R. Brumbaugh. 2012. Historical ecology with real numbers: Past and present extent and biomass of an imperilled estuarine habitat. Proceedings of the Royal Society B: Biological Sciences, doi:10.1098/rspb.2012.0313.

Appendix A

Major U.S. Coastal Storms Since 1900

Table A-1 provides information on major coastal storms that have impacted the United States since 1900. The committee primarily utilized information compiled by NOAA's National Hurricane Center and the National Weather Service and included storms from those compilations that were reported to have caused over $200 million in damages (in 2013 dollars) or over 200 deaths, although the list is not exhaustive. Most of the storms included on this list are tropical cyclones, which may be due in part to the relatively localized geographic impacts of these storms and the focus of post-event analyses. Extratropical storms, exemplified by two notable storms in this list, tend to affect much larger areas of the coast and produce waves and surges that usually persist for much longer than tropical systems. For this reason, they produce much more destruction to natural coastal defenses (primarily the dune system and their fronting beaches). The lack of careful post-event analyses of these storms makes it very difficult to find accurate estimates of the total damages; however, it should be recognized that they occur much more frequently than tropical cyclones along most East Coast areas and play a large role in changing the vulnerability and resilience of coastal communities.

The Saffir-Simpson scale is included in Table A-1 only to provide a concept of storm intensity as the storms struck the U.S. coast. Many other factors influence storm surge and coastal damages. The Saffir-Simpson scale was abandoned by the National Hurricane Center as an indicator of storm surge/coastal inundation in 2010 (http://www.nhc.noaa.gov/).

TABLE A-1 Major U.S. Coastal Storms since 1900

Name	Year	Location	Estimated U.S. Damages (in Billions of 2013 Dollars)	Fatalities	Saffir-Simpson Category*
Hurricane Sandy[a]	2012	Florida to Maine	66	159	2
Hurricane Irene[a]	2011	Puerto Rico, NC, mid-Atlantic coast, New York City, NY	14	45	1
Hurricane Ike[a]	2008	Galveston Island, Texas	32	112	2
Hurricane Wilma[a]	2005	Naples, FL; Upper Keys, FL; Marathon, FL	23	35	3
Hurricane Rita[a]	2005	Texas, Louisiana	22	119	3
Hurricane Katrina[a]	2005	Buras, LA	149	1833	3
Hurricane Dennis[a]	2005	Gulf coast, FL	3	15	3
Hurricane Jeanne[a]	2004	Puerto Rico, Florida	9	28	3
Hurricane Ivan[a]	2004	Southeastern U.S.	25	57	4
Hurricane Frances[a]	2004	Florida	12	48	2
Hurricane Charley[a]	2004	Florida, New Jersey	20	35	4
Hurricane Isabel[a]	2003	Mid-Atlantic	7	55	2
Tropical Storm Allison[a]	2001	Texas, North Carolina	11	43	Tropical storm
Hurricane Floyd[a]	1999	North Carolina	9	77	2
Hurricane Opal[a]	1995	Florida	5	27	3
Hurricane Andrew[a]	1992	Lower east FL coast; Gulf Coast	45	61	4
Hurricane Iniki [a]	1992	Hawaii	5	7	4
The "Perfect Storm"[b]	1991	Florida through Maine	<1	5	Extratropical storm
Hurricane Hugo[a]	1989	Puerto Rico; Charleston, SC; Hatteras, NC	17	86	4
Hurricane Alicia[a]	1983	Galveston, TX	7	21	3
Tropical Storm Claudette[c]	1979	Texas, Louisiana, Oklahoma	1	2	Tropical storm
Hurricane Agnes[c]	1972	East coast of Florida, Pennsylvania, New York	12	122	1

TABLE A-1 Continued

Name	Year	Location	Estimated U.S. Damages (in Billions of 2013 Dollars)	Fatalities	Saffir-Simpson Category*
Hurricane Camille[c]	1969	Gulf Coast	9	256	5
Ash Wednesday Storm[d]	1962	East coast from Cape Hatteras, NC to Rhode Island	1.5	40	Extratropical Storm
Hurricane Donna[c]	1960	Puerto Rico, Florida, North Carolina, New England	3	50	4
Hurricane Audrey[c]	1957	Texas, Louisiana	1	390	4
Hurricanes Connie and Diane[c]	1955	North Carolina	7.6	184	3 (Connie), 1 (Diane)
Hurricane Hazel[c]	1954	South Carolina, North Carolina	2	95	4
Hurricane Carol[c]	1954	North Carolina, Virginia, New York	4	60	3
Great Atlantic Hurricane[c]	1944	North Carolina to Maine	1.3	46	3
New England Hurricane[c]	1938	North Carolina, New York, Connecticut	5	600	3
Florida Keys Labor Day Hurricane[c]	1935	South Florida	<1	408	2
San Felipe-Lake Okeechobee Hurricane[c]	1928	Puerto Rico, Florida	<1	2148	4
Great Miami Hurricane[c]	1926	Miami, FL	91	373	4
Atlantic Gulf Hurricane[c]	1919	Florida, Texas	<1	600-900	4
Galveston Hurricane[c]	1900	Galveston, TX	<1	6,000-8,000	4

NOTES: This list is not exhaustive. *Saffir-Simpson category provided at landfall on the U.S. coast.

SOURCES:
[a]http://www.ncdc.noaa.gov/billions/events,
[b]http://www.ncdc.noaa.gov/oa/satellite/satelliteseye/cyclones/pfctstorm91/pfctstdam.html,
[c]http://www.nhc.noaa.gov/outreach/history/,
[d]http://www.erh.noaa.gov/er/lwx/Historic_Events/StormsOfCentury.html.

Appendix B

USACE Coastal Storm Damage Reduction Projects

TABLE B-1 USACE Coastal Storm Damage Reduction Projects

Project Name	Project State	Initial Construction Date	Project Type	Project Length (miles)
Mobile County—Dauphin Island Sand Pilot	AL	2011	Beachfill	20
Gulf Beach, Milford	CT	1957	Beachfill	0.23
Prospect Beach, West Haven	CT	1957	Beachfill	1.1
Sea Bluff Beach, West Haven	CT	1991	Beachfill	0.19
Sherwood Island State Beach, Westport	CT	1957	Beachfill	1.5
Southport Beach, Fairfield	CT	1958	Beachfill	0.13
Woodmont Beach, Milford	CT	1995	Beachfill, Groin	0.38
Stamford Hurricane Barrier	CT	1969	Storm Surge Barrier	2
New London Hurricane Barrier	CT	1986	Storm Surge Barrier, Levees	0.61
Pawcatuck Hurricane Barrier	CT	1964	Storm Surge Barrier, Levees	0.35
Point Beach	CT	2004	Nontructural	NA
Gulf Street, Milford	CT	1987	Revetment	0.03

continued

TABLE B-1 Continued

Project Name	Project State	Initial Construction Date	Project Type	Project Length (miles)
Middle Beach, Madison	CT	1957	Revetment	0.13
Bridgeport (Port V)	CT	1984	Revetment	0.05
Delaware Bay Coastline: Roosevelt Inlet—Lewes Beach	DE	2004	Beachfill	0.3
Delaware Coast, Cape Henlopen to Fenwick Island: Bethany—South Bethany	DE	2008	Beachfill	2.8
Delaware Coast, Cape Henlopen to Fenwick Island: Fenwick Island	DE	2005	Beachfill	1.2
Delaware Coast, Cape Henlopen to Fenwick Island: Rehoboth Beach—Dewey Beach	DE	2006	Beachfill	2.6
Delaware Coast Protection, Indian River Inlet Sand Bypassing	DE	1989	Sand Bypassing	0.5
North Shore Indian River Inlet	DE	1988	Revetment	0.3
South Shore Indian River Inlet	DE	1988	Revetment	0.3
Brevard County—North Reach	FL	2000	Beachfill	9
Brevard County—South Reach	FL	2002	Beachfill	3
Broward County SPP—Segment II (Ft. Lauderdale)	FL	1979	Beachfill	11
Broward County SPP—Segment III (Hollywood/Hallandale)	FL	1988	Beachfill	8
Dade County BEC—Bal Harbor	FL	1975	Beachfill	11
Dade County BEC—Sunny Isles	FL	1988	Beachfill	3
Duval County BEC	FL	1978	Beachfill	10

TABLE B-1 Continued

Project Name	Project State	Initial Construction Date	Project Type	Project Length (miles)
Fort Pierce Beach SPP	FL	1970	Beachfill	1
Lee County BEC—Captiva	FL	1988	Beachfill	5
Lee County BEC—Gasparilla	FL	2007	Beachfill	3
Manatee County SPP—Anna Maria Island	FL	1992	Beachfill	4
Martin County HSDR	FL	1995	Beachfill	4
Nassau County SPP	FL	2008	Beachfill	4.3
Palm Beach SPP—Delray Beach	FL	1973	Beachfill	3
Palm Beach SPP—Jupiter/ Carlin	FL	1995	Beachfill	1.1
Palm Beach SPP—North Boca Raton	FL	1988	Beachfill	1
Palm Beach SPP—Ocean Ridge	FL	1997	Beachfill	1
Panama City Beaches	FL	2010	Beachfill	16.3
Pinellas County—Long Key	FL	1980	Beachfill	4
Pinellas County—Sand Key	FL	1988	Beachfill	8
Pinellas County—Treasure Island	FL	1969	Beachfill	4
Sarasota County—Venice Beach	FL	1997	Beachfill	3
St. Johns County BEC	FL	2001	Beachfill	3
Virginia Key	FL	2000	Groins	2
Tybee Island	GA	1975	Beachfill	3.5
Grand Isle and Vicinity	LA	1985	Dune with geotube core, stone jetty, offshore breakwaters)	7.5
Lake Pontchartrain and Vicinity	LA	1965	levees, floodwalls, floodgates, surge barriers	65

continued

TABLE B-1 Continued

Project Name	Project State	Initial Construction Date	Project Type	Project Length (miles)
West Bank and Vicinity—Resilient Features	LA	2013	Levees, Floodwalls	15.4
Salisbury Beach	MA	2011	Beachfill	0.23
Newburyport Beach	MA	2011	Beachfill	0.44
North Scituate Beach, Scituate	MA	1967	Beachfill	0.47
Plum Island Beach, Newbury	MA	1973	Beachfill	0.17
Revere Beach	MA	1991	Beachfill	2.5
Roughans Point, Revere	MA	1997	Revetment	0.59
Quincy Shore Beach	MA	1959	Beachfill, Bulkheads	1.6
Clark Point Beach, New Bedford	MA	1980	Beachfill,Groins	0.3
Oak Bluffs Town Beach	MA	1973	Beachfill,Groins	0.23
Town Beach, Plymouth	MA	1968	Beachfill,Groins	0.03
Wessagusset Beach, Weymouth	MA	1959	Beachfill,Groins	0.49
Winthrop Beach	MA	1959	Beachfill,Groins	0.8
Charles River Dam	MA	1978	Storm Surge Barrier	0.08
New Bedford Hurricane Barrier	MA	1966	Storm Surge Barrier	3.41
Town River Bay, Quincy	MA	1992	Revetment	0.05
Bluffs Community Center	MA	1994	Revetment	0.06
Island Ave	MA	1983	Revetment	0.05
Point Shirley	MA	1995	Revetment	0.15
Assateague Island Restoration—Short Term & LTSM	MD	2002	Ecosystem Restoration	5
Atlantic Coast MD Storm Protection (Ocean City)	MD	1990	Beachfill, Seawall	8.9
Alley Bay, Beals	ME	1979	Revetment	0.1
Holmes Bay, Whiting	ME	1980	Revetment	0.21

TABLE B-1 Continued

Project Name	Project State	Initial Construction Date	Project Type	Project Length (miles)
Islesboro (The Narrows)	ME	1984	Revetment	0.06
Johnson Bay, Lubec	ME	1980	Revetment	0.08
Machias Bay, Machiasport	ME	1994	Revetment	0.05
Marginal Way, Ogunquit	ME	1987	Revetment	0.05
Roosevelt Campobello International Park, Lubec	ME	1989	Revetment	0.07
Sand Cove, Gouldsboro	ME	1984	Revetment	0.1
Merriconeag Sound, Harpswell	ME	1979	Seawall	0.05
Hancock County Beaches	MS	2005	Beachfill	5.3
Comprehensive Barrier Island Restoration	MS	2011	Ecosystem Restoration	8.3
Harrison County—Deer Island Ecosystem Restoration—I	MS	2010	Beachfill, Ecosystem Restoration	4
Harrison County Beach Dunes	MS	2010	Beachfill	24
Jackson County—Pascagoula Beach Ecosystem Restoration	MS	2009	Beachfill	1.4
Hancock County—Bay St Louis Seawall	MS	2010	Beachfill, Seawall	1.6
Hancock County—Bayou Caddy Shoreline Protection	MS	2010	Breakwater, Ecosystem Restoration	0.4
Brunswick County Beaches (Ocean Isle Beach)	NC	2001	Beachfill	18
CAP—Section 1135 (Sea Turtle Habitat Project, Oak Island)	NC	2001	Beachfill	1.7
Carolina Beach and Vicinity, Area South (Kure Beach)	NC	1998	Beachfill	3.4
Carolina Beach and Vicinity, Carolina Beach Portion	NC	1965	Beachfill	2.7
Wrightsville Beach	NC	1965	Beachfill	2.7

continued

TABLE B-1 Continued

Project Name	Project State	Initial Construction Date	Project Type	Project Length (miles)
Fort Macon	NC	1834	Groins	1.5
Fort Fisher	NC	1996	Revetment	0.6
Hampton Beach, Hampton	NH	1955	Beachfill, Groins	1.2
Wallis Sands State Beach, Rye	NH	1983	Beachfill, Groins	0.2
Ocean Gate	NJ	2002	Beachfill	0.8
Brigantine Island	NJ	2005	Beachfill	1.4
Cape May City (Cape May Inlet to Lower Township)	NJ	1989	Beachfill	3.6
Keansburg	NJ	1968	Beachfill	2.8
Laurence Harbor	NJ	1965	Beachfill	1.9
Lower Cape May Meadows—Cape May Point	NJ	2004	Beachfill	4.4
Ocean City (Great Egg Harbor Inlet and Peck Beach)	NJ	1991	Beachfill	4.5
Sea Bright—Manasquan: Asbury to Avon	NJ	1999	Beachfill	3
Sea Bright—Manasquan: Belmar to Manasquan	NJ	1997	Beachfill	6
Sea Bright—Manasquan: Long Branch	NJ	1997	Beachfill	3
Sea Bright—Manasquan: Monmouth Beach	NJ	1994	Beachfill	3
Sea Bright—Manasquan: Sea Bright	NJ	1995	Beachfill	3
Absecon Island (Atlantic City and Ventnor)	NJ	2004	Beachfill	8.5
Barnegat Inlet to Little Egg Inlet (LBI)	NJ	2012	Beachfill	18
Townsends Inlet—Cape May Inlet	NJ	2002	Beachfill, Seawalls	4.3
East Point	NJ	2012	Revetment	0.1
Coney Island	NY	1993	Beachfill	3

TABLE B-1 Continued

Project Name	Project State	Initial Construction Date	Project Type	Project Length (miles)
East Rockaway Inlet to Rockaway Inlet and East Rockaway Inlet to Rockaway Inlet Section 934	NY	1975	Beachfill	6.2
Fire Island Inlet to Shores Westerly	NY	1973	Beachfill	2.7
Orchard Beach	NY	2010	Beachfill	1
West of Shinnecock Inlet	NY	2004	Beachfill	0.8
Westhampton	NY	1996	Beachfill	4
Oakwood Beach	NY	2000	Levee, storm surge barrier	0.1
Orient Harbor	NY	2011	Revetment	0.11
Shelter Island	NY	1999	Revetment, Bulkhead	0.19
Village of Northport	NY	2004	Revetment, Bulkhead	0.02
Misquamicut Beach, Westerly	RI	1960	Beachfill	0.62
Oakland Beach, Warwick	RI	1981	Beachfill, Groins, Revetment	0.04
Fox Point Hurricane Barrier, Providence	RI	1966	Storm Surge Barrier	0.55
Cliff Walk	RI	1972	Revetment, Walkways	3.4
Folly Beach	SC	1992	Beachfill	5.3
Hunting Island	SC	2002	Beachfill	0.5
Myrtle Beach Reach 1—North Myrtle Beach	SC	1997	Beachfill	8.6
Myrtle Beach Reach 2—Myrtle Beach	SC	1997	Beachfill	9
Myrtle Beach Reach 3—Garden City/Surfside	SC	1998	Beachfill	8
Galveston Seawall	TX	1902	Seawall	10
Sargent Beach Revetment	TX	1998	Revetment	8

continued

TABLE B-1 Continued

Project Name	Project State	Initial Construction Date	Project Type	Project Length (miles)
Chesapeake Bay Shoreline, Hampton	VA	2005	Beachfill	0.7
Virginia Beach Hurricane Protection	VA	2001	Beachfill	6
Wallops Island	VA	2012	Beachfill	4.5
Sandbridge Beach	VA	2002	Beachfill	0.5
Jamestown Island Seawall	VA	1969	Seawall	0.3
Cape Charles Shore Protection	VA	1992	Seawall	0.1
Norfolk Floodwall	VA	1971	Floodwall	0.5
Anderson Park Shore Protection	VA	1979	Revetment	0.3
Hampton Institute Shore Protection	VA	1976	Revetment	0.3
Tangier Island Shore Protection	VA	N/A	Revetment	1.1
Saxis Island Bulkhead	VA	1989	Bulkhead	0.1

SOURCE: Donald Cresitello, USACE, personal communication, 2014.

Appendix C

Biographical Sketches of Committee Members

Richard A. Luettich, Jr., *Chair*, is the Sewell Family Term Professor of Marine Sciences and Director of the Institute of Marine Sciences at the University of North Carolina (UNC), Chapel Hill. Dr. Luettich also serves as director of the UNC Center for Natural Hazards and Disasters in Chapel Hill and is the lead principal investigator on the Department of Homeland Security Center of Excellence in Natural Disasters, Coastal Infrastructure and Emergency Management. His research deals with modeling and measurement of circulation and transport in coastal waters. Dr. Luettich's modeling efforts have emphasized the development and application of unstructured grid solution techniques for geometrically complex systems such as sounds, estuaries, inlets, and inundated regions. He codeveloped the ADCIRC coastal circulation and storm surge model that has been applied extensively for modeling storm surge along the U.S. coast. Dr. Luettich has also participated in the development of components of the national Coastal Ocean Observing System. He served on the National Research Council committees to review the Louisiana Coastal Protection and Restoration Program and the New Orleans Hurricane Protection System. He received his B.S. and M.S. degrees in civil engineering at the Georgia Institute of Technology and his Sc.D. in civil engineering from the Massachusetts Institute of Technology.

Gregory B. Baecher, NAE, is the Glenn L. Martin Institute Professor of Engineering at the University of Maryland. His research focuses on the reliability of civil infrastructure and risks posed by natural hazards and the response of infrastructure to those hazards. In recent years, his research has dealt with dam safety and with the response of levee systems to flooding, including actuarial issues related to flood and other natural hazard insurance. He has also worked on quantitative methods in facilities management, especially federally owned facilities, and on information technology applications to facilities management. Dr. Baecher was elected to the National Academy of Engineering in 2006 for his work in the development, explication, and implementation of probabilistic- and reliability-based approaches to geotechnical and water resources engineering. He is a recipient of the Commander's Award for Public Service from the U.S. Army Corps of Engineers and a recipient of the Thomas A. Middlebrooks Award and State-of-the-Art Award from the American Society of Civil Engineers. He is coauthor of *Reliability and Statistics in Geotechnical Engineering* (2003), *Risk and Uncertainty in Dam Safety* (2004), and *Protection of Civil Infrastructure from Acts of Terrorism* (2006). Dr. Baecher received his Ph.D. and M.Sc. degrees in civil engineering from the Massachusetts Institute of Technology and his B.S. in civil engineering from the University of California, Berkeley.

Susan S. Bell is professor of Marine Ecology in the Department of Integrative Biology at the University of South Florida (USF). Dr. Bell's research focuses on topics in marine ecology, especially landscape ecology of marine systems, restoration ecology, and marine conservation. Many of her ongoing studies target questions related to ecosystem response to changing marine habitats. Her work focuses on seagrass habitats (quantifying large-scale distribution and change) but includes investigations in other coastal areas including mangroves, salt marshes, and sandy beaches. In addition, Dr. Bell collaborates with a group of researchers, mainly based at USF, who are working on issues linking urban ecology, watersheds, and human dimensions. Dr. Bell received a Ph.D. in 1979 from University of South Carolina.

Phillip R. Berke is professor in the Department of Landscape Architecture and Urban Planning, and director of the Institute for Sustainable Coastal Communities, Texas A&M University at College Station. He is also a collaborative research scholar of the International Global Change Institute in New Zealand, and a faculty affiliate with the Plan Evaluation Lab of the University of British Columbia. Dr. Berke's research interests include land-use and environmental planning, state and local development management, sustainable development, and natural hazard mitigation in developed and developing communities. His research seeks to explore the causes of land use decisions and the consequences to the environmental, social, and economic systems of human settlements. He was a member of the Science and Engineering Board for the 2012 Update of Louisiana's Master Plan for Coastal Protection and Restoration and a member of the NRC's Committee on Disaster Research in the Social Sciences: Future

Challenges and Opportunities. Dr. Berke received his B.A. in economics and environmental science from Empire State College, M.S. in natural resources planning from the University of Vermont, and Ph.D. in urban and regional science from Texas A&M University.

Ross B. Corotis, NAE, is the Denver Business Challenge Professor of Engineering at the University of Colorado (UC), Boulder. His research interests are in the application of probabilistic concepts and decision perceptions for civil engineering problems, with particular focus on societal tradeoffs for hazards in the built infrastructure. His current research emphasizes the coordinated roles of engineering and social science with respect to framing and communicating societal investments for long-term risks and resiliency. He previously served on the faculty at Northwestern University, established the Department of Civil Engineering at the Johns Hopkins University—where he was also associate dean of engineering—and was dean of the College of Engineering and Applied Science at UC Boulder. He has numerous research, teaching, and service awards, was editor of the *International Journal Structural Safety* and the *Journal of Engineering Mechanics*, and chaired the Executive Committee of the International Association for Structural Safety and Reliability. He is a member of the National Academies Board on Infrastructure and the Constructed Environment and previously served on the Disasters Roundtable Steering Committee and the Committee on Integrating Dam and Levee Safety and Community Resilience. Dr. Corotis was elected to the National Academy of Engineering in 2002. He received his S.B., S.M., and Ph.D. in civil engineering from the Massachusetts Institute of Technology.

Daniel T. Cox is professor in the Coastal and Ocean Engineering Program and adjunct faculty of the College of Earth, Ocean, and Atmospheric Sciences at Oregon State University. Before coming to OSU in 2002, he was associate professor of civil engineering at Texas A&M University. His research focuses on coastal processes, particularly nearshore hydrodynamics, sediment transport, surf-zone turbulence, and boundary-layer processes. He also has an interest in the design and performance of coastal structures. Dr. Cox is an associate editor for the *Coastal Engineering Journal* and a member of the American Society of Civil Engineers' Subcommittee to Develop Standards for Tsunami Engineering Design. He received his Ph.D., M.S, and B.S. degrees in civil engineering from the University of Delaware.

Robert A. Dalrymple, NAE, is the Willard and Lillian Hackerman Professor of Civil Engineering at Johns Hopkins University in Baltimore, Maryland. His major research interests are in the areas of coastal engineering, wave mechanics, fluid mechanics, littoral processes, and tidal inlets. His research currently explores water wave modeling, tsunamis and their impacts on shorelines, and the interaction of water waves with the seabed, specifically mud bottoms. Dr. Dalrymple was elected to the National Academy of Engineering in 2006. He chaired the NRC Committee on the Review of the Louisiana Coastal Protection and Restoration Program and the NRC Committee on Sea Level Rise in California, Oregon, and Washington. Dr. Dalrymple received his A.B. degree in engineering sciences from Dartmouth University, his M.S. degree in ocean engineering from the University of Hawaii, and his Ph.D. degree in civil and coastal engineering from the University of Florida.

Tony MacDonald is currently the director of the Urban Coast Institute at Monmouth University, West Long Branch, New Jersey. Mr. MacDonald was previously the executive director of the Coastal States Organization from 1998 to 2005. Prior to joining the Coastal States Organization, he was the special counsel and director of environmental affairs at the American Association of Port Authorities, where he represented the International Association of Ports and Harbors at the International Maritime Organization on negotiations on the London Convention. He has also practiced law with a private firm in Washington, D.C., working on environmental and legislative issues, and served as the Washington, D.C. environmental legislative representative of the Mayor of the City of New York. He specializes in environment, coastal, marine, and natural resources law and policy and federal, state, and local government affairs. He earned a B.A. from Middlebury College and a J.D. from Fordham University.

Karl F. Nordstrom is a professor in the Institute of Marine and Coastal Sciences at Rutgers University. His research is focused on the dynamic processes affecting the size, shape, and location of beaches and dunes in ocean and estuarine environments. His research also includes analysis of natural hazards, land use, and restoration of naturally functioning environments in developed municipalities. He has worked in the United States, Canada, Australia, Italy, and Germany, and has published numerous books, including *Beaches and Dunes of Developed Coasts* and *Estuarine Shores: Evolution, Environments, and Human Alterations*. He received Fulbright Senior Scholar Awards in 1999 and 2006, and the Grove Karl Gilbert Award for Excellence in Geomorphological Research. He is on the editorial board of the *Journal of Coastal Research* and is a member of several professional associations on coastal environments and beach preservation. Dr. Nordstrom received his M.S. and Ph.D. in geography from Rutgers University.

Stephen Polasky, NAS, is the Fesler-Lampert Professor of Ecological/Environmental Economics at the University of Minnesota. He previously held faculty positions at Oregon State University and Boston College. Dr. Polasky was also the senior staff economist for environment and resources for the President's Council of Economic Advisers from 1998 to 1999. His research interests include ecosystem services, natural capital, biodiversity conservation, endangered species policy, integrating ecological and economic analysis, renewable energy, environmental regulation, and common property resources. He has served as coeditor and associate editor for the *Journal of Environmental Economics and Management*, as associate editor for the *International Journal of Business and Economics*, and is currently serving as an associate editor for *Conservation Letters*, *Ecology and Society*, and *Ecology Letters*. He was elected to the National Academy of Sciences in 2010. He is also a fellow of the American Academy of Arts and Sciences and the American Association for the Advancement of Science. He received his Ph.D. degree in economics from the University of Michigan.

Sean P. Powers is professor and chair of Marine Sciences, University of South Alabama, and senior marine scientist, Dauphin Island Sea Lab. Dr. Powers' research focuses on the ecology of coastal and estuarine fishes and benthic invertebrates, particularly those that support commercial and recreational fisheries. His current research includes efforts to quantify the linkages between habitats (natural, restored, and constructed) and demersal fishes and invertebrates, conservation and restoration of marine biogenic habitats, and development of ecosystem-based management approaches. Much of Dr. Powers' research is focused on the interface of social, economic, and ecological sciences and how this interaction influences sustainable management of natural resources. Dr. Powers received his B.S. in biology and chemistry from Loyola University, an M.S. in biological sciences from the University of New Orleans, and a Ph.D. in biology, with areas of specialization in ecology and evolution, zoology, and biostatistics, from Texas A&M. He currently serves as a committee member for the Science and Statistical Committee for the Gulf of Mexico Fishery Management Council and as a scientific advisor for the National Oceanic and Atmospheric Administration's Natural Resource Damage Assessment for the *Deepwater Horizon* Oil Spill.

Don Resio is professor of ocean engineering and director of the Taylor Engineering Research Institute at the University of North Florida, where he is building a new advanced degree program and developing a new curriculum in Coastal and Estuarine Engineering. He is a recognized leader in meteorology, hydrodynamics, and probabilistic analysis of environmental hazards in coastal, estuarine, and riverine areas. Dr. Resio's research interests include the development of innovative marine and coastal structures, environmental statistics (with a focus on weather extremes), surface gravity waves in deep and shallow water, improved wave measurement systems, and coastal processes. Dr. Resio previously served as the senior technologist for the U.S. Army Corps of Engineers Coastal and Hydraulics Lab from 1994 to 2011. He served as a coleader of the post-Katrina interagency forensics analysis of wave and storm surge effects on levees and subsequently became the leader of the risk analysis team for the South Louisiana Hurricane Protection Project, including consideration of the effects of climatic variability on hurricane characteristics in the Gulf of Mexico. This team developed a new technical approach for hurricane risk assessment now being used along all U.S. coastlines, which is also being extended by the Nuclear Regulatory Commission for new licensing guidelines at coastal sites. Dr. Resio currently serves as a U.S. delegate to the United Nations' Joint World Meteorological Organization's Technical Commission for Oceanography and Marine Meteorology (JCOMM) in the area of climate effects in the ocean and is the cochair of the UN Coastal Inundation and Flooding Demonstration Project. Dr. Resio earned his Ph.D at the University of Virginia in environmental sciences.

Ap Van Dongeren is a senior researcher at Deltares in the Netherlands. His research interests include wave generation, nearshore circulation, and nearshore morphology and dune erosion. Dr. Van Dongeren has been project leader on a number of national and international projects, including development and application of the Delft3D model for the Office of Naval Research. He has also led the Deltares effort to develop the open-source dune erosion model XBeach. He has led a project team to improve the performance of SWAN (a wave model) in order to derive more reliable wave boundary conditions for flood risk assessments. He is the research program leader on event-driven hydro- and morphodynamics, and is the coordinator of a European Union project on Resilience-Increasing Strategies for Coasts. Dr. Van Dongeren received his M.Sc. from Delft University of Technology and his Ph.D. in coastal engineering from the University of Delaware.